The Laws of Connection

Also by David Robson

The Intelligence Trap
The Expectation Effect

The Laws of Connection

The Scientific Secrets of Building a Strong Social Network

DAVID ROBSON

PEGASUS BOOKS
NEW YORK LONDON

THE LAWS OF CONNECTION

Pegasus Books, Ltd.
148 West 37th Street, 13th Floor
New York, NY 10018

First Pegasus Books cloth edition June 2024

ISBN: 978-1-63936-648-4

10 9 8 7 6 5 4 3 2

Printed in the United States of America
Distributed by Simon & Schuster
www.pegasusbooks.com

To Robert

CONTENTS

INTRODUCTION

Looking back on her first years, Helen Keller saw her deafblindness as a kind of captivity. Without any means of relating to others, she was a prisoner of her own mind, and her increasingly desperate family could not find a way for her to escape.

Her liberation would come in the form of a twenty-year-old woman named Anne Sullivan, who taught Keller how to communicate through touch to the skin. It was an incredible release from her solitary confinement. 'Love came and set my soul free,' she later wrote. 'Once I fretted and beat myself against the wall that shut me in . . . But a little word from the fingers of another fell into my hand that clutched at emptiness, and my heart leapt to the rapture of living.'[1]

With language, she could finally share her thoughts and feelings; simply learning the names of everyday objects increased her 'kinship with the rest of the world'. On discovering the word 'love', she experienced a 'beautiful truth' bursting into her mind. 'I felt that there were invisible lines stretched between my spirit and the spirits of others.'[2]

Keller would grow up to weave a rich social network, including a surprisingly close bond with the writer Mark Twain, who described her as the 'fellow to Caesar, Alexander, Napoleon, Homer, Shakespeare, and the rest of the immortals'.[3] Her autobiography is as much a celebration of friendship as an account of her extraordinary tenacity. 'My friends have made the story of my life,' she noted in its final pages. 'In a thousand ways they have turned my limitations into beautiful privileges and enabled me to walk serene and happy in the shadow cast by my deprivation.'[4] Keller repeated this sentiment throughout her life, memorably declaring that, 'I would rather walk with a friend in the dark than walk alone in the light.'

Few of us can imagine the isolation that Keller must have experienced before she met Anne Sullivan. But I'm sure we can all identify, at least to a small degree, with the feeling of being divided from others, and the exhilaration of finally finding an affinity with someone else and experiencing the tug of those 'invisible lines' that tie us to the people we love. The craving for connection is a universal experience. If I track the emotional landscape of my own life, I find that loneliness accompanied me into the deepest valleys, while moments of mutual understanding pushed me to its highest peaks.

We now know that deep bonds do not just bring deep pleasure. According to multiple studies, a greater sense of connection is consistently linked to greater health and longevity. Social activities reduce psychological distress, protect us from infection, and lower the risk of Alzheimer's and heart disease.[5] When people feel that they have strong social support, they also perform better on tests of problem-solving and creativity, and enjoy greater professional success.[6]

Given the sheer wealth of evidence, scientists are evangelical about the power of connection to improve our wellbeing. Yet few of us are getting enough of it. The most common measure of loneliness asks participants to indicate how often they lack companionship, feel excluded, or have the sense that 'there is no one I can turn to'. Using this scale, research over many years has shown that around 50 to 60 per cent of US citizens report feeling social disconnection at regular intervals in their lives.[7]

One explanation is that changes to the structure of modern society have made it harder for us to meet people. Commentators have repeatedly pointed to the fact that we are less likely to live near our relatives, and that digital technology is taking us away from face-to-face meetings. This cannot be the whole story, however. Historical accounts suggest that loneliness has been a major preoccupation for decades, and people who work in big offices, have large families and are invited to glamorous parties still feel underappreciated and unloved.[8] For most people it is the absence of a close emotional bond with the people around us that leaves us feeling disconnected and isolated, rather than a mere lack of social opportunities.

Cutting-edge scientific research can help us to understand why that is. According to an exciting new theory, a deep sense of connection comes from constructing a 'shared reality' with another person. Put simply, this is the knowledge that the other person thinks and feels and interprets events in broadly the same way as us – that they understand us, have the same visceral feelings, and experience the world in the same way. When a shared reality is established between two people, their neural activity begins to synchronise; their interactions flow more smoothly, they feel greater trust and affection, and their stress levels plummet.

At its most glorious extreme, the forging of a shared reality between two people creates the sensation that their minds have merged with one another – as, indeed, Keller described when writing about her friendship with Anne Sullivan. 'My teacher is so near to me that I scarcely think of myself apart from her,' she wrote. 'I feel that her being is inseparable from my own, and the footsteps of my life are in hers.'[9] When we lack a shared reality with the people around us, however, we feel literally alienated – as if we are a foreign visitor speaking an entirely different language.

In this book, I will be drawing on groundbreaking new evidence to show you the ways that we construct a shared reality with the people that we meet – and the common psychological barriers that can prevent it from forming. Every time we engage with another person, we make decisions that could either bring greater understanding and affection, or continued distance and isolation. As a result of our brains' biases, we all too often choose the latter path – accidentally sabotaging the chance to build a shared reality.

Discovering these common errors can make for unsettling reading, but this research also offers exciting opportunities. By learning to identify and overcome the psychological barriers to connection, we can all establish more meaningful relationships with the people around us – strangers, colleagues, friends, siblings, parents and romantic partners. You do not need to be blessed with natural charisma, charm or effortless self-confidence. Through a change of mindset, anyone can start to create a more fulfilling and stable social network, with all the benefits for our health and creativity that come with it.

For a flavour of this research, just consider a recently discovered phenomenon known as the 'liking gap', which leads us to ignore the potential for connection – even when it is staring us in the face.

Like many scientific discoveries, the finding was inspired by personal experience. A few years ago, the psychologist Erica Boothby happened to be engaged in a conversation with a new acquaintance, while her partner – another psychologist, Gus Cooney – stood close by. After the conversation, Boothby was worried that she'd made a bad impression. To Cooney's ear, however, the exchange had been warm and friendly. What could possibly have gone wrong?

As the two psychologists chatted about the event, they began to wonder whether this is a common human experience – that after a meeting, we consistently underestimate how much the other person enjoyed our company. We lose faith in the shared reality created in the first meeting, and our doubts weaken the bond that had formed.

Boothby and Cooney named the phenomenon the 'liking gap' – and set about investigating its prevalence. In the first study, participants were placed into pairs and given a five-minute ice-breaking task, before answering questionnaires about how much they liked the other person, how much they thought the other person liked them, and whether they would like to see each other again. As the researchers had hypothesised, most people were overly pessimistic about the impression that they'd given, as they started to question whether they'd really established a mutual understanding. In general, the other person liked them a lot more, and would have been much keener to follow up with another meeting, than they ever imagined.

The liking gap is the reason we may enthusiastically exchange numbers or emails with a new acquaintance – but then never send a message or make a call. Boothby and Cooney's initial findings have been replicated many times, with one experiment showing that the liking gap can linger through months of regular contact with people. Despite living together, university roommates continued to feel insecure about the ways their companions viewed them for the best part of a year, for instance. Later research has shown that the liking gap is also prevalent among colleagues in the workplace, where it can limit creative collaborations.

When I first read about the liking gap, I couldn't help but flinch while thinking about all the times I might have ignored overtures of friendship. As a science writer specialising in psychology and neuroscience, however, I was deeply excited. Since the 1970s, behavioural economists such as Daniel Kahneman have been outlining the many cognitive biases that lead our financial decisions astray. Now we seemed to be witnessing the birth of a whole subfield of social psychology that could do exactly the same for our relationships.

I wasn't wrong. Over the past few years, social psychologists have written a torrent of papers outlining the many judgement errors that prevent us from connecting with other people and the ways to overcome them. These new findings cover everything from our fear of making new acquaintances to the complexities of navigating disagreement and conflict.

For example, we vastly overestimate how awkward it will be to talk to someone we don't know, but people are often more than grateful to strike up a conversation with a stranger – with huge benefits for everyone's wellbeing. And when we do have the chance to connect, we shy away from discussing deeper topics in favour of superficial small talk, yet it is exactly the more profound conversation that would foster the creation of a shared reality. Equally importantly, our compliments and our apologies are often woefully misjudged, so that we fail to say the words that would help us to reinforce or repair the mutual understanding that is essential for a strong bond. And our fears of seeming needy or incompetent prevent us from asking for help, when a simple request for assistance can increase our standing in others' eyes. Indeed, asking for a favour is counterintuitively one of the best ways to build a rapport with someone, improving both your and their wellbeing – a phenomenon psychologists call the Benjamin Franklin Effect (we will find out why later in the book).

These are just a few of the ways our intuitions prevent us from establishing the mutual understanding that contributes to more meaningful relationships. From more than 300 academic papers, I have gleaned thirteen overarching principles that will help you to build a more satisfying social network. I call these the laws of connection.

If you already feel socially confident, you may wonder if these findings would be of benefit to you, or whether they are aimed at the overtly shy or socially awkward. But the research suggests that the majority of people, across all personality types, need to correct these biases if they are to enjoy a more meaningful social life. Even the most gregarious people can suffer from misguided intuitions that are unwittingly pushing others away.

I should emphasise that this is high-quality research – a fact that cannot always be taken for granted when discussing psychological discoveries. In this field (and indeed many other areas of science) some eye-catching studies have withered under later scrutiny, thanks to small numbers of participants and poorly designed experiments – resulting in a so-called 'replication crisis'. Such concerns are not a problem here. Keenly aware of these past mistakes, the social psychologists behind the latest findings have been scrupulous in their methods, verifying their ideas on hundreds and often thousands of people in diverse circumstances, so that we can be as sure as possible that the conclusions reflect true phenomena.

Needless to say, the application of this research does require a certain degree of diligence. We must always be certain to consider the context of our actions and respect others' boundaries. (I'll elaborate more on these points, where relevant, in later chapters.) Provided that you proceed with sensitivity and respect, however, this research should be cause for great optimism. We can all build more rewarding and fulfilling relationships – through practical steps that are possible for anyone to apply.

If you're feeling a little sceptical about what you can achieve, I'd like to share my own story and the reasons that I have come to write this book.

As a child and teenager, I was so shy that I would find myself tongue-tied whenever I met a stranger. My dream was to be a journalist, but the very idea that I could have a career interviewing people about their research seemed laughable. As I approached the time to go to university, however, I decided it was time to overcome my shyness, and so, step by step, I forced myself to overcome my own 'liking gap'. I learned to recognise that people are often far more willing to build connections than my intuitions had led me to believe. My fears of hostile reactions were almost always unfounded, and even when I made the odd blunder, people were far more forgiving than I had imagined. Through persistent effort, my social confidence became great enough for me to embark on my career – and I have never looked back.

When I came across the recent research on all the psychological barriers to social connection, I deeply regretted that I hadn't known these facts earlier in my life. *The Laws of Connection* is the book that I wish I could have read whenever I lacked confidence and needed a guide to navigate new social challenges. Whether you are a shy teen going to college, an expat embarking on a new life in a foreign land, or simply one of the many people who would like to build better bonds with the people that you know, this book is for you. While this is decidedly not a dating or parenting manual, you may find that the tips can supercharge your love life or help you to grow better relationships with your family; they can also ease your interactions with colleagues at work. The laws of connection are rooted in our human nature and apply to every person we meet.

Part 1 examines the neuroscience and physiology of social bonds. You'll learn how two minds can meld into a shared sense of reality, and why that brings such astonishing benefits for our health and wellbeing. You'll also discover tools to assess the state of your current social network – including ways to identify the 'ambivalent relationships' (or 'frenemies') that might be causing you more harm than good. We'll then explore how to build new connections. I'll explode the myth that certain personality types, such as shy people, will find it inherently difficult to form deep relationships. You'll learn the conversational traps such as the 'novelty penalty' and the 'illusions of understanding' that prevent us from establishing rapport – and the ways to avoid them.

Part 2 is all about maintaining and nourishing bonds for lifelong connection. We'll examine inevitable difficulties that might arise in any relationship, and the ways to heal fractures in our shared reality. Are lies ever justified, or should we always act with 100 per cent honesty? What is the secret to giving negative feedback? How should you ask for help without losing respect? How can you celebrate your successes without inviting jealousy or resentment? And what's the secret of an effective apology? In each case, our intuitions are often wildly off the mark – but the theory of shared reality can offer much-needed guidance for the ways to tackle these challenges.

Finally, in the conclusion, we'll go digital to examine the future of friendship in the metaverse and beyond. Dismantling the hackneyed claims about the dangers of social media, we'll discover the specific ways that new technologies can strengthen or damage relationships, and we'll learn how to apply the principles of authentic social connection through any medium. This will provide us with our final law of connection.

Writing these pages, I've certainly extended and deepened my existing friendships and made new ones along the way – and I hope that by reading the chapters that follow, you will too.

PART 1

BUILDING CONNECTIONS

CHAPTER 1

THE SOCIAL CURE

In 1981, Yossi Ghinsberg made the greatest mistake of his life. A naive twenty-two-year-old Israeli, fresh out of military service, he had embarked on a backpacking trip across South America to La Paz, Bolivia. Along the way, he acquired new friends, including Karl Ruprechter, an Austrian 'geologist' who promised to take the group deep into the Amazon, through previously unchartered territory.

Ruprechter proved to be little more than a con artist who could not even swim.[1] As the tensions in the group grew, Ghinsberg and another member of the group, Kevin Gale, decided to break away from the others, by taking a raft along the Tuichi River. After a few hours, however, they entered white-water rapids. The raft crashed into a rock, and Gale managed to swim to the banks while Ghinsberg was washed downriver and over a waterfall.

What followed is one of the most astonishing survival stories in recent history. In his three-week journey back to civilisation, Ghinsberg faced venomous snakes, wild jaguars, and parasitic worms that burrowed deep into his skin. That's not to mention the extreme hunger that led him to spend hours at a time daydreaming about all the food he would eat after he had been rescued.

It was the lack of human company, however, that threatened to break Ghinsberg. 'I suffered most from the loneliness,' he later wrote. A staunch individualist, he remembered once feeling scornful of the idea that 'people need people'. Now he saw that social connection was as essential as food or water. He even created imaginary friends to stave off mental collapse.[2]

Such feelings are common to anyone who has suffered extreme isolation. Just consider the words of the late US senator John McCain, describing his experience of solitary confinement as a prisoner of war in Vietnam: 'It crushes your spirit and weakens your resistance more effectively than any other form of mistreatment . . . The onset of despair is immediate, and it is a formidable foe.' Finding covert ways to communicate with the other prisoners was 'a matter of life and death', he said.[3]

Evolution wouldn't have programmed us to feel this anguish at being alone if relationships were not one of life's fundamental requirements. According to an enormous body of research, social connection is as essential for our long-term health as a balanced diet and regular physical activity, while loneliness can be a slow-acting poison that severely reduces our lifespan. Social connection can also boost our creativity and productivity, and it even comes with unexpected professional benefits – all of which should make our life less stressful and more rewarding.

C.S. Lewis was precisely wrong when he wrote that 'Friendship is unnecessary . . . It has no survival value; rather it is one of those things which gives value to survival.'[4] And if we want to learn how to build better social connections, we should first understand why they are so important.

THE ALAMEDA 8

Let's first examine the evidence behind the health benefits. In the early 1960s, Lester Breslow at the California State Department of Public Health set out on an ambitious project to identify the habits and behaviours that led to greater longevity. To do so, he recruited nearly 7,000 participants from the surrounding Alameda County. Through detailed questionnaires, he built an extraordinarily detailed picture of their lifestyles, and then tracked their wellbeing over the subsequent years.

Within a decade, Breslow's team had identified many of the ingredients that we now know are essential for good health: don't smoke; drink in moderation; sleep seven to eight hours a night; exercise; avoid snacks; maintain a moderate weight; eat breakfast. At the time, the findings were so striking that when his colleagues presented the results to him, he believed they were playing some kind of prank.

You will hardly need me to explain these guidelines in more detail – the 'Alameda 7' are now the basis of most public health guidance.[5] The research continued, however, and by 1979, two of Breslow's colleagues – Lisa Berkman and Leonard Syme – had discovered an eighth factor that influenced people's longevity: social connection. On average, the people with the greatest number of ties were around half as likely to die as the people who had smaller networks.[6] The result remained even after they had controlled for factors such as socioeconomic status, the quality of people's health at the start of the survey, and those other health practices that had been proven to be so important for longevity, such as cigarette consumption, exercise and diet.

Delving deeper, it became clear that all kinds of relationships mattered, but some were more meaningful than others. A sense of connection with spouses and close friends offered the greatest protection, but even casual acquaintances at church or a bowling club helped to stave off the grim reaper.[7]

The sheer audacity of the claim may explain why it was initially neglected in public health guidance; scientists were used to seeing the body as a kind of machine, largely detached from our mental state and our social environment. But extensive research building on the Alameda study now confirms that connection and loneliness influence our susceptibility to many diverse diseases.[8]

Social support can boost your immune system and protect you from infection, for instance. In the late 1990s, Sheldon Cohen at Carnegie Mellon University in the US asked 276 participants to give full details of their social ties. They were tested for an existing infection, then placed in quarantine and asked to inhale water droplets laced with rhinovirus – the bug behind many coughs and sneezes. Over the following five days, many of the participants went on to develop symptoms, but this was significantly less likely if they had a large and diverse range of social connections. Indeed, those with the lowest levels of social connection had three to four times the risk of developing a cold than those with richer networks of family, friends, colleagues and acquaintances.

Any good scientist should always consider whether other confounding factors might explain the result. It's reasonable to assume that lonely people could be less fit and active, for example, if they spend less time out and about with friends and family. As Berkman and Syme had also discovered, however, the link remained even after the researchers accounted for all those factors in their statistical analyses.[9] And the size of the effect vastly exceeds the benefits of popping vitamin supplements – another measure we might take to boost our immune system.[10]

The social health boost extends to our risk of chronic, life-changing conditions such as type 2 diabetes. This arises when the pancreas stops producing enough insulin, and the body's cells stop responding to the insulin that's flowing through the blood – both of which prevent it from breaking down blood sugar to power cells. Factors such as obesity can contribute to diabetes, and so, it seems, does the quality of your social ties. A study of 4,000 participants in the English Longitudinal Study of Ageing research found that people's scores on the UCLA Loneliness Scale predicted the onset of type 2 diabetes over the following decade.[11] There are even some good signs that people with stronger social connections have a reduced risk of developing Alzheimer's disease and other forms of dementia.[12]

The strongest evidence, however, concerns cardiovascular diseases, with massive longitudinal studies tracking the health of tens of thousands of people repeatedly highlighting the link. This can be seen in the earliest stages – people with poor social relationships are more likely to develop hypertension – and in the worst outcomes, with loneliness increasing the risk of a heart attack or stroke by about 30 per cent.[13]

To get a measure of the social health boost's overall importance, Julianne Holt-Lunstad at Brigham Young University compiled the findings of 148 studies, covering 300,000 participants, that had looked at the benefits of social integration and the hazard of social disconnection. She then compared the effects of loneliness with the risks of various other lifestyle factors, including smoking, drinking alcohol, exercise and physical activity, body mass index (a measure of obesity), air pollution and taking medication to control blood pressure.

The results, published in 2010, were truly astonishing: Holt-Lunstad found that the size and quality of people's social relationships either equalled or outmatched almost *all* the other factors in determining people's mortality.[14] The more people feel supported by the people around them, the better their health and the less likely they were to die. Overall, social connection – or its absence – played a larger role in people's health than alcohol consumption, exercise, body mass index and air pollution. Only the effects of smoking came close.

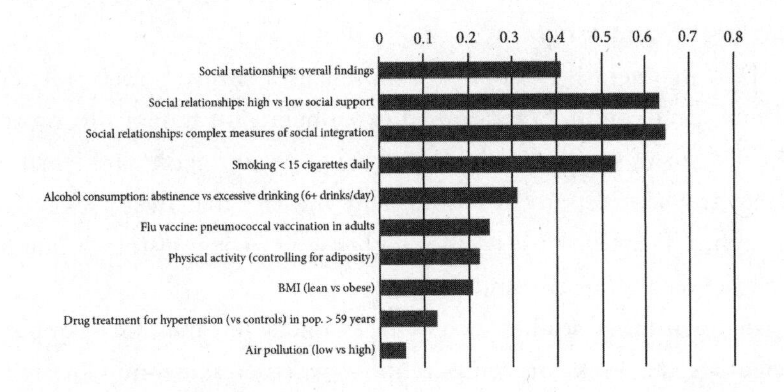

Above you can see a graph summarising Holt-Lunstad's findings. To interpret it, you need a little knowledge of 'effect size', which measures, on a scale of 0 to 1, a lifestyle factor's practical importance. The larger the effect size, the stronger its relationship to longevity. As you can see, the different elements of social connection have larger effect sizes than many commonly accepted predictors of good health. Social integration reflects the structure and size of someone's social networks as well as their participation in social activities, while social support reflects perceptions of loneliness and care from the people around them. Both prove to be essential for a long life.

This research has faced critics. For iron-clad proof of a causal link between one lifestyle factor and overall longevity, you'd need to conduct a controlled experiment, in which you randomly allocate people to different conditions. That's how drugs are tested – some take the pill and others take a placebo, and you then record the different outcomes. In this case, you would have to allocate some people to a loneliness condition, denying them friendships, while others are given a ready-made social network full of loving people. Clearly, this is impossible – a fact that has led some people to question whether the apparent effects of social connection are real and significant. They suggest that, despite the scientists' best efforts, they might have missed some confounding factor that gives the illusion of a link between our social lives and our health and longevity.

This argument isn't quite as damning as it seems, however. After all, we can't conduct randomised experiments on humans to prove the dangers of smoking – the ethics would be even more problematic – but few scientists today would deny the fact that the one causes the other. That's because there are other ways to demonstrate a causal link between a lifestyle and a disease.

In longitudinal studies such as the Alameda research, for example, scientists can look for 'temporality' – whether someone's lifestyle choice precedes the development of illness. In this case, the sequence is very clear: the people reported their loneliness long before they developed their ill health. Scientists can also look for a 'dose–response relationship' – whether greater exposure of the proposed lifestyle factor results in greater risk. Once again, there is a clear pattern – someone who is completely isolated is more likely to suffer worse health than someone who is occasionally lonely, who in turn suffers more illness than someone who has a vibrant social circle. You can also check whether the findings are consistent across different populations and using different measurement types. If the effects have only been identified in one small sample, or if they only appear when you consider a single loneliness questionnaire, you would be right to be sceptical. But as we have already seen, this is not the case.

The social health boost has now been documented across the world using multiple methods to quantify people's social connectedness. Whether you are asking for subjective feelings or considering objective facts, such as someone's marital status or the exact number of times they see acquaintances each month, the pattern stays the same.[15] We can even see parallel effects in other social species as diverse as dolphins, chacma baboons and rhesus macaques: the more integrated an individual is within its group, the greater its longevity.[16] It is for all these reasons that the majority of scientists studying public health believe that social connection is one of the key determinants of health and longevity.

SAFETY IN NUMBERS

To understand *how* and *why* the strength of our social ties could influence our health to such a degree, we must consider our evolution. As humans adapted to living in bigger and bigger groups, everything from our supply of food to protection from predators would have depended on our relationships with others. To lose our standing with our companions would have left us in danger of starvation and illness and injury.

As a result, the brain and body have evolved to interpret social isolation as a serious threat. This is the reason we feel such anguish when we are lonely and disconnected. In much the same way that physical pain warns us to seek out safety and tend to our wounds, social pain should persuade us to avoid hostile parties and re-establish our social ties.[17]

Feelings of rejection or seclusion also set off a cascade of physiological reactions. In our evolutionary past, these were supposed to protect us from the immediate danger that might result from our isolation, such as attacks from predators or enemies. The brain triggers the release of norepinephrine and cortisol – hormones that keep our minds alert to threats and prepare the body for aggression. The immune system, meanwhile, starts ramping up the production of inflammatory molecules. Inflammation provides a first line of defence from pathogens, which would reduce our risk of infection if we happened to be injured by an attack.[18] A sense of isolation and social stress can also increase the creation of fibrinogen. This promotes blood clotting and would have helped protect us against dangerous blood loss if we suffered a wound.[19]

While this response would have increased our chances of short-term survival, it can cause longer-term damage. When the body is constantly prepared for hostility and aggression, it puts extra strain on the cardiovascular system. Chronic inflammation, meanwhile, may prevent our wounds from being infected, but the accompanying immune response is less adept at responding to viruses – which would increase the chances that we might get infected with a respiratory illness, for example.[20] Chronic inflammation also causes wear and tear on our other cells that can raise the risk of diabetes, Alzheimer's and heart disease. The elevated levels of the clotting factor fibrinogen, meanwhile, can cause thrombosis, which may lead to a heart attack or stroke.

If we spend decades in loneliness and isolation, these changes can drastically raise the risk of illness and early death. When people enjoy connection and social support, however, their bodies will suppress processes such as inflammation. As a result, they will have a much better baseline of health that renders them less susceptible to disease.[21]

Connection may be so important for social animals that we appear to have special 'loneliness neurons', which become more and more active when we are alone. In much the same way that hunger or thirst drives us to seek food and water, they produce a craving that pushes us to reach out to others and to bask in their company. Once we feel sated, we may be happier with time alone – until those loneliness neurons become active again.[22]

If we have lacked connection for too long, we may simply stop listening to that signal, in much the same way that someone with an eating disorder might try to disregard their physical hunger. With the laws of connection, however, we can look for long-term remedies that will help us to build meaningful relationships that contribute to better health and wellbeing.

UNLEARNING FEAR

Besides reducing the direct physiological consequences of loneliness, social connections create a kind of psychological buffer from life's other stresses.

When people face a task such as public speaking, those with knowledge of greater social support tend to show more muted changes to their blood pressure and levels of the hormone cortisol, suggesting that their body is reacting more calmly to the challenge.[23] Perhaps unconsciously, the brain has assessed all the resources available to it, and knows that the consequences of a failure would not be so drastic if you have others watching your back and ready to provide security and comfort if you need it, and this prevents it from going into full fight-or-flight mode.

Reminders of social support also help people to erase fear memories, meaning that negative associations are less likely to linger in their minds after they are no longer relevant. In the laboratory, you can test the formation of fear memories with a standard experimental procedure, in which certain pictures or symbols are accompanied by a small electric shock. Soon, the very sight of the image puts us into a state of stress – and that lingers even when the scientists stop delivering the shock. Research by Naomi Eisenberger at the University of California, Los Angeles shows that those fear memories are much weaker, and disappear more quickly, when people are also primed with reminders of people they love.[24]

Social support can even ease bodily pain. We saw earlier how loneliness can create the illusion of pain, but this is the exact opposite effect: it's the reduction of discomfort. Simply looking at a photo of a loved one increases people's tolerance of a painfully hot probe placed on the skin, for instance.[25] Pain is known to be influenced by our perceptions of safety: it's partly a signal to hide away and tend our wounds, rather than risking further injury. And if we have someone close by – or are reminded of their presence with a photo – we don't need such a strong signal, since they can provide the necessary care and protect us from further danger.

By buffering stress and extinguishing fear memories, social support can reduce the emotional baggage of negative experiences, so that we can recover from trauma more quickly. This might explain why social support is so protective against a range of mental illnesses. When veterans return from war, for instance, the strength and quality of their social ties offers one of the best predictors of whether they will go on to develop post-traumatic stress disorder.[26] And in the enormous strain of the Covid-19 pandemic, people with the most robust social connections were the least likely to suffer from poor psychological health.[27]

CONNECTED CREATIVITY

The final mechanism for the social health boost comes from the material and cognitive benefits of having high social capital. When we are unemployed, for instance, friends and friends of friends often provide a route back into work, which is one reason why people with stronger ties have greater financial stability. A study of unemployment in Spain found, extraordinarily, that 84 per cent of people rely on their existing contacts to find a new position. In the US, that figure is around 66 per cent, and in the UK, 50 per cent.[28]

Once we have landed a job, our social connections can contribute to greater productivity and creativity. The hypothesis here is primarily one of cross-pollination. If you know and like many different people, you are more likely to come across multiple perspectives that could teach you something interesting about your own job. (And in return, you might provide a valuable insight that is useful for theirs.) Your contacts can provide useful feedback that will help to refine your thoughts, and to help spread the word once your ideas are ready to be disseminated.

Historical analyses provide circumstantial evidence for this proposition. Dean Simonton at the University of California, Davis carefully examined the biographies of 2,026 scientists and inventors. He first ranked their 'eminence' based on the number of times they were featured in various scientific reference works, and the length of their entries – and then compared this with the number of social relationships revealed in the available record, including their relatives, intimate friends, mentors, collaborators, and pupils. Simonton's analyses revealed a clear correlation between the total number of these connections and the scientists' eminence.

Isaac Newton exemplified the finding. Of all the 2,026 scientists, he was considered to be the most eminent, and one of the most well connected. In the popular imagination, Newton is often thought to be the epitome of the solitary genius, working in near confinement. William Wordsworth, for instance, described Newton as 'Voyaging through strange seas of thought, alone' – but it seems that he had plenty of passengers accompanying him on his intellectual journeys. His correspondents included his rival Gottfried Wilhelm Leibniz; the philosopher John Locke, who, despite finding Newton to be a prickly character, noted that he had 'several reasons to think him truly my friend'; and the astronomer Edmund Halley, who encouraged Newton in the writing of his monumental *Principia Mathematica* and even helped to fund its publication.

'The notion of the "lone genius" must be myth if the merits of even the most acclaimed creators partly reflect the assets of the society in which they reside,' concludes Simonton.[29]

The power of creative cross-pollination gave rise to Broadway's most iconic musicals. Management scientists have examined the professional networks behind hundreds of New York's top shows from the twentieth century. The analyses focused on six specialist roles – the composer, the lyricist, the librettist, the choreographer, the director, and the producer – for each production. The researchers first determined how frequently each of the team members had previously worked together, and then compared that to the financial and critical success of the shows that the team created.

Some teams were extremely tight-knit – the same people would work together again and again without inviting any fresh faces. Their insular relationships resulted in more stale productions, which were considerably more likely to fail at the box office. The teams behind the most successful shows, in contrast, included newcomers and members who had previously worked with other groups. These well-connected 'creative butterflies' carried fresh ideas and new approaches between the different groups – all of which helped to attract larger audiences to their shows. *West Side Story* is the perfect example. The team included lyricist Stephen Sondheim, a relative newcomer, alongside composer Leonard Bernstein, director Jerome Robbins and his assistant choreographer Peter Gennaro – who had ample experience between them, but who had never worked together before. The mix of outlooks resulted in one of the most iconic musicals of all time.[30]

While we must be a little cautious in interpreting these historical analyses, more recent surveys reveal a similar pattern. People at the centre of wide social networks tend to come up with more original ideas, and capitalise on them, than those with fewer ties in a more restricted network.[31]

In business, the increased sharing of ideas and information through social networks makes it more likely that you can profit from new opportunities and markets – a fact that became abundantly apparent after the fall of the Berlin Wall in 1989. Entrepreneurs with a greater number of social ties crossing the former divide between East and West Germany were more likely to find business opportunities in, and profit from, the vast economic developments across the country. Importantly, this income boost was not driven by existing business partnerships; for decades, the economic divide between East and West had severed professional networks spanning the two regions. Instead, it was the social links between families that seemed to make the difference.[32]

If nothing else, a capacity for connection makes our work more satisfying – a fact that should not be overlooked, given that the average person spends roughly a third of their waking time dealing with their colleagues. Across multiple industries – from construction work to medicine and education – the most reliable predictors of job satisfaction are frequent interaction with others, the appreciation of office friendships, and the emotional support that they supply.[33] A lack of connection at work, meanwhile, brings a much higher risk of burnout.[34]All of these advantages – from finding work to building our careers and enjoying our success with like-minded colleagues – would contribute to a less stressful life, offering yet more reasons why we see such a strong wellness premium for people with strong social ties.

The essential benefits of social connection for our health and happiness are now so well established that it should be a priority for every one of us – but the message does not yet seem to have reached the general public. While the seven other factors identified by the Alameda study – such as exercise, diet, and seven to eight hours' sleep a night – have been the fodder for many health campaigns, most people underestimate the role of their social lives in determining their health. When researchers asked 500 British and American participants to estimate the importance of all the various health behaviours that Holt-Lunstad included in her meta-analysis, social integration and social support came near the very bottom of the public's rankings. This study was published in 2018, decades after the link first became known, and despite the hundreds of studies showing that the social health boost equals or even exceeds the impact of those other lifestyle factors.[35]

More than forty years after Berkman and Syme made their discovery, it's time for opinions to change. As the authors of one recent study put it, we need to recognise that 'connection is medicine'.[36] Building and maintaining our social ties is just as important as having a new gym membership, eating our five fruit and veg a day or getting our vaccinations. It should be a priority for anyone who values their time on earth.

THE BEST OF FRENEMIES

Just as your nutritional requirements must be tailored to your size and body type, the ideal social diet will vary from person to person. In general, people who live with someone they love tend to have greater wellbeing than those who live alone, and people who socialise with friends, relatives or colleagues around once a week are healthier and happier than those who have face-to-face contact less than once a month.[37] Beyond that, there are no hard-and-fast rules for the optimum size of a social network or the frequency of social interaction that will apply to every person.

In any case, the quality of our relationships will matter as much as the quantity, with a growing recognition that certain connections may do us more harm than good.

To get a flavour of this research, pick a couple of people within your social network and answer the following questions on a scale of 1 (not at all) to 6 (very much). When you are feeling in need of advice, understanding or a favour:

- How helpful is your connection?
- How upsetting is your connection?

The scale was developed by Julianne Holt-Lunstad and colleagues at the University of Utah, who had also conducted the large academic review examining the overall importance of social connection for our health – and they have used it to create three broad categories of relationships.[38]

I hope that the majority of your connections scored highly on the first question, while getting the lowest possible rating on the second. These are your *supportive* social ties. For a few, however, your answers would be the mirror image – they are typically hurtful and barely ever helpful. These are purely *aversive* relationships: the people you'll do your best to avoid talking to unless you are forced to interact, in a business meeting or at a family gathering, for example. You'll also have a few who score 1 on both questions. These are your indifferent relationships – a neighbour, perhaps, who is rather bland company with neither good nor bad qualities.

A handful of your connections, however, may represent something of a paradox. You may rate them as 4 on helpfulness and 4 on hurtfulness. Anyone with a score of 2 or more on both scales is considered an *ambivalent* connection. I know I have at least a couple: an acquaintance who can be incredibly generous yet also lashes out with a bitter putdown when she feels envious or threatened, and a former colleague who helped me through many work crises but occasionally claimed credit for my best ideas, sometimes without even acknowledging my contributions. We might commonly call these 'frenemies', but they could also be a parent or a sibling. And the ambivalence can come in many forms: it might be lack of interest in your life rather than overt dis-respect, or a general unreliability that means they're often unavailable when you expect them to be watching your back. You might have an ambivalent spouse who 'love-bombs' you one day but who is fiercely critical the next, leaving you unsure about their true feelings.

Ambivalent connections deserve particular attention, since they have unique consequences for our health. Holt-Lunstad's team, for example, hooked 102 people up to portable cardiovascular monitors for three days. During each social interaction, the participants could press a button to trigger a reading, and after they'd finished the conversation, they recorded who they'd met and rated them on the scales above. This allowed the researchers to see the different stress responses triggered by each kind of relationship.

As you might expect, people's blood pressure was higher when they met an ambivalent tie compared to when they met someone who was uncomplicatedly supportive. Surprisingly, however, the ambivalent ties also provoked a stronger reaction than the purely aversive people. Something about the uncertainty of the interaction made it much more stressful than meeting someone who was reliably unkind.[39]

Later studies confirmed the effect. Holt-Lunstad found that simply knowing that an ambivalent connection was in the next room as they completed another task was enough to send people's blood pressure rocketing.[40] And whereas reminders of uncomplicatedly positive friends or family members tend to soothe us, the merest mention of an ambivalent connection – presented as a flash on a computer screen – can trigger a stress response.[41]

It seems that our ambivalent connections have us in a kind of stranglehold. We may depend on their support, and try our best to please them, but that emotional investment means that an ambivalent connection's occasional nastiness will be especially hurtful. And the uncertainty about which side of them we are going to see – the Dr Jekyll or the Mr Hyde – only compounds the stress of meeting them, so that we feel anxious before they have even opened their mouths.

If they play too much of a role in your life, the long-term effects of ambivalent connections may be just as bad as having few connections at all. Regularly interacting with ambivalent connections can put extra strain on the heart and raises levels of bodily inflammation, which, as we've seen, can put you at risk of a range of different diseases.[42]

The effects can even be seen in measures of cellular ageing. At the end of our chromosomes, we have protective caps called telomeres that prevent our DNA from being damaged when cells replicate. With the stresses of life, our telomeres slowly wear down, and when they become too short the cell may start to malfunction or die. Shorter telomeres are thought to put us at greater risk of many of the diseases that come from ageing – and our ambivalent connections appear to contribute to their decline. If you live with someone who often makes you feel like you are living on a knife edge, or if you regularly see friends who leave you feeling that way, you are more likely to have shortened telomeres, relative to other people of a similar age.[43]

Ambivalent connections can be common in the workplace, and if those people are in a position of power, their unreliable support and sporadic unkindness can take a toll on their employees' mental health, raising the risk of depression, anxiety and exhaustion.[44]

The paradoxical nature of our ambivalent connections means that there is no simple solution. If you feel that they have become too toxic a presence in your life, you may decide to cut off contact – but that's not possible if they are your boss or a family member, or if they are so deeply integrated into your social network that you would also risk other supportive relationships.

Simply being aware of the ambivalent nature of a relationship might offer some protection, however. Personally speaking, I have found that knowledge of this research allows me to mentally prepare myself for the mixed feelings that my own ambivalent connections might bring. This allows me to focus more on the good in the relationships and to feel compassion for their more unpleasant streaks, while also attempting to reduce contact when I feel that they may add to the stresses that I'm facing in other areas of my life.

Just as importantly, this research has prompted me to look at my own behaviour to ensure that I was not proving to be an ambivalent connection myself. Inevitably, I found that I was. I was guilty of not showing enough interest or pleasure in another's successes; not giving their opinions and experiences enough respect; always waiting for the other person to contact me first and then not responding quickly enough – and with enough enthusiasm – to their messages; or forgetting to message them on their birthdays. This leads us to our first law of connection: **Be consistent in your treatment of others. Avoid being a stressful frenemy.**

Over the following chapters, we'll learn more about these behaviours and many others that can weaken our social ties – and the ways to strengthen those bonds. We may not be able to turn every ambivalent connection into a positive relationship, but we can ensure that we act in the most supportive ways ourselves, and in many cases this will help us to bring out the best in others too.

* * *

Before we move on, you might be wondering what happened to Yossi Ghinsberg. In the decades since his isolation in the Amazon, he has continued to travel and has worked on multiple philanthropic projects in his home country of Israel, the US and Bolivia – the location of his misadventure.

Having once dismissed the idea that 'people need people', he now places social connection at the centre of his life philosophy. 'With the right mindset we can focus on the similarities that connect us rather than the differences, a realization that brings with it a sense of purpose and responsibility,' he writes. 'It is my core belief that separation is an illusion that we can choose not to cultivate.'[45]

Ghinsberg claims that his plight in the jungle was the first time he realised how much he needed the company of others.[46] Given the growing scientific literature on the huge benefits of social connection to our health and happiness, we might all benefit from a similar wake-up call.

What you need to know

- Social connection is one of the most important predictors of physical and mental health. It soothes pain, reduces inflammation, and reduces the risk of thrombosis
- Even small reminders of loved ones – such as a photo – can reduce our startle response and soothe upsetting memories
- People with a greater number of social connections are more creative and have greater professional opportunities, which leads to more financial security
- The size of our social network and the frequency of interactions matter, but most important is the quality of the relationships, which can be supportive, aversive or ambivalent. The supportive relationships are the most beneficial, and the ambivalent ones are the most harmful

Action points

- Try to identify the ambivalent relationships in your social network and think of the ways they make you feel and influence your behaviour. If they are the ambivalent frenemies, look for ways to decompress after your meetings
- Question whether you ever act ambivalently to other people, and ask what you might do to change your behaviour to show how much you value them

CHAPTER 2

HOW WE CONNECT

Returning to Smith College after Thanksgiving in 1950, eighteen-year-old Sylvia Plath felt profoundly disconnected from everyone around her. On first joining the college, she had eagerly written to her mother about the wonderful feelings of camaraderie with her housemates. She said she had never known such friendliness.[1]

After her visit home for the holidays, however, the thought of meeting the other girls, with their 'artificial chatter', filled Plath with despair. She saw an almost insurmountable gulf between her own mind, fully occupied with recollections and dreams and sensory impressions, and the inner lives of the other students, concealed behind their 'false grinning faces' and their 'tinsel gaiety'. In her journal she described her loneliness as being 'a disease of the blood'.

Could this void – between the self and other – ever be crossed? In that moment, Plath sounded sceptical. 'Life is loneliness,' she wrote. 'And when at last you find someone to whom you feel you can pour out your soul, you stop in shock at the words you utter – they are so rusty, so ugly, so meaningless, and feeble from being kept in the small cramped dark inside you for so long.'[2]

Plath's loneliness was very different from that of the foolhardy backpacker Yossi Ghinsberg, whom we met in the last chapter. He was physically isolated from other people; she was living in halls populated by many other students who, at least superficially, should have been just like her. But as many of us have felt, simply having others around is not enough. When we crave company, we want to feel truly understood; we want to know that other people are thinking, feeling and perceiving the world in the same way that we are.

Psychologists call this the experience of having a 'shared reality', and they have shown that it is the foundation upon which all meaningful social connection is built.[3] In our most profound relationships, we may even feel that the other person is a part of ourselves – a line of thinking that can be traced back to Aristotle, who described friendship as having 'one soul dwelling in two bodies'.[4] When we lack that feeling of having a shared reality with the people around us – as Plath did in her dorm room the night after Thanksgiving – we are said to experience 'existential isolation'.

By recognising the ways that shared reality can be formed and broken in the brain, we can start to understand the basic principles of building deep and authentic bonds. This not only inspires our second law of connection – **Create a mutual understanding with the people you meet; ignore superficial similarities and instead focus on your internal worlds, and the peculiar ways that your thoughts and feelings coincide** – it also underlies many of the other lessons from the rest of this book.

BIRDS OF A FEATHER

Folklorists have long suspected that 'birds of a feather flock together'. The recent research on shared reality, however, has a very specific prediction: that it is the sharing of internal states – thoughts, feelings and perceptions – that creates social connection, over and above surface-level similarities in background or circumstance. Put simply, we want to know that someone is experiencing the world in the same way as we are.

Imagine, for instance, that you are on an induction day at a new office and you find out the following facts about your new colleague:

- You both giggle at the same joke
- You both cry at the same song
- You both feel awe at the same artwork

- You both come from the same hometown
- You both attended the same university

How would these facts influence your perception of that person?

According to theories of shared reality, it is the first three statements that should make the biggest difference to our first impressions, since they reflect something immediate about our inner lives. And that is exactly what psychological research shows: our liking for someone immediately jumps when we find that they have had the same reactions to an event – however trivial. Other facts about someone's background, such as whether they come from the same hometown, do help to create a rapport, but they are generally less powerful than the knowledge of their thoughts and feelings.[5]

Some of the most compelling evidence comes from Elizabeth Pinel at the University of Vermont, who has spent two decades exploring existential isolation and the kinds of experiences that help us to escape it. In one study, she asked participants to play a version of the board game Imaginiff, in which you have to consider various absurd scenarios, such as: *Imagine if Jennifer Aniston were a tool. Would she be a cocktail mixer, a screwdriver, a sledgehammer, or a pair of toenail clippers?* In line with the shared reality theory, Pinel found that people had a much stronger liking for someone who chose similar responses to themselves, and they were more willing to work with them later.

Why should we care whether someone sees Jennifer Aniston as a screwdriver or a pair of toenail clippers? The result only makes sense if we believe those answers can tell us something about the workings of our mind, and if someone's responses are the same as our own, it suggests that they are processing the world in the same way that we are. Pinel describes this as 'I-sharing', since it reveals something immediate about your subjective experience of the world, a common 'state of consciousness'. And when it came to forming a sense of connection, I-sharing proved to be more important than some seemingly obvious identity markers, such as the participants' sexual orientation, which might have provoked an in-group or out-group prejudice. Pinel found similar results with gender and even race.[6]

Pinel's latest research has shown that perceiving an intimate shared reality can bridge political divides. In the days running up to the fiercely contested 2020 US presidential election, she asked 417 Donald Trump and Joe Biden supporters, of diverse ages and backgrounds, to individually interpret a series of images of inkblots. You can see an example below:

What do you see when you look at the inkblot?

A. a monster
B. a clown
C. a jack-o-lantern
D. a smile

After they had made their choices, participants were told some (sham) feedback about what another person had perceived. They were then asked to rate how much they would like that other person, how close they felt to them, and how much they wanted to be friends with them.

Unsurprisingly, during this febrile time in US politics, the Trump and Biden supporters generally did not hold high opinions of each other; all other things being equal, they preferred people of their own political stripe. If they discovered that their partner had shared the same interpretation of the inkblot, however, they felt considerably warmer towards that person, regardless of their differing political beliefs.[7] Much like the answers to the Imaginiff questions, the interpretations of the inkblots are fairly meaningless in themselves. But the discovery that their partner saw the images in the same way nevertheless gave the participants the *perception* that their two minds were somehow attuned – and this created a temporary sense of connection.

Pinel has now replicated and extended these findings many times. She has shown, for example, that sharing knowledge about people's inner states can increase generosity and trust. In laboratory games, for example, participants were happier to share a greater portion of their potential winnings if they experienced a sense of shared reality. We don't just like someone more if we know that they think and feel the same way as us; we are also more likely to help them, even if it comes at a personal cost.[8]

MERGED MINDS

The creation of shared reality may help us to overcome the frailties of human perception. In any situation, the data that the brain receives is very ambiguous, and there are many ways that it can interpret what it sees and hears. In isolation, we have no way of knowing whether our understanding of events is valid. When we know that someone else shares our thoughts and feelings, however, it provides the necessary reassurance that our experiences are valid, allowing us to have greater faith in our own judgements. Knowing that someone else experiences the world in the same way that we do will also make the other person's behaviours more predictable, which will also create greater feelings of security.

Pinel has found that priming people with feelings of existential isolation renders them particularly sensitive to signs of shared reality. She asked people to read the following instructions and then write down their thoughts: 'You can be lonelier in a crowd than by yourself. With this saying in mind, please now think of a situation in your past when you felt disconnected or very isolated from the other people around you.' Afterwards, these participants were asked to declare their liking for different people, and they proved to be far more easily swayed by information about the other's inner life, compared to participants who had written about a time they felt really bored. Having experienced that profound sense of loneliness, they were actively looking for validation of their thoughts and feelings.[9]

Each of our relationships will be the product of hundreds or thousands of tiny events that either underline or undermine our sense of shared reality. It could be that you both opt for the same meal at a restaurant, spontaneously express joy at the same sunset, or feel similar schadenfreude at a colleague or relative's mishap. Perhaps you have a habit of saying the same things at the same time or finishing each other's sentences. Each of these things will reinforce your sense that you are experiencing the world in the same way.

No two people will ever agree on everything, but in general, the stronger the perception that you share the same inner life, the more connected you will feel to another person. Psychologists at Texas A&M University asked nearly 300 people in romantic couples to answer: 'How often do you have moments where you and your partner feel the same way in response to something you experience together?' The participants' responses robustly predicted how much they believed that their partner understood their 'true' selves, and their overall relationship satisfaction.[10]

If you are wondering how much you experience shared reality with your friends, colleagues and relatives, you can also use a questionnaire designed by Maya Rossignac-Milon, then at Columbia University. Simply think of an individual within your life, and rate the following statements on a scale of 1 (strongly disagree) to 7 (strongly agree):

1. We frequently think of things at the exact same time.
2. Through our discussions, we often develop a joint perspective.
3. We typically share the same thoughts and feelings about things.
4. Events feel more real when we experience them together.
5. The way we think has become more similar over time.
6. We anticipate what the other is about to say.
7. We are more certain of the way we perceive things when we are together.
8. We often feel like we have created our own reality.

This is the 'Generalised Shared Reality Scale' or SR-G, and the individual answers are averaged to give a final score. Your position on this spectrum won't necessarily reflect how outwardly similar you are to the person in question, or even how much you esteem them. But it will almost certainly reveal how much you like them and how close you feel to them – as Rossignac-Milon and her colleagues demonstrated in multiple studies.[11]

In their first experiment, the team gathered more than 600 people who had been in a straight or gay couple for an average of nine years and gave them detailed questionnaires about the quality of their relationship. Most people arrived at an SR-G score of around 5, but there was huge variation, with some getting close to 1 and others scoring a perfect 7. As expected, Rossignac-Milon found that the higher scores on the SR-G predicted greater trust, commitment and satisfaction with each other.

To broaden her sample, Rossignac-Milon next recruited a further 545 participants and asked them to complete daily records about their interactions with a significant person in their life – either a romantic partner, friend, roommate, parent, or sibling. The diaries included the SR-G scale and the Inclusion-of-Other-in-Self scale, a simple test in which participants are asked to state which of the following pictures best describes their feelings for the other person.[12]

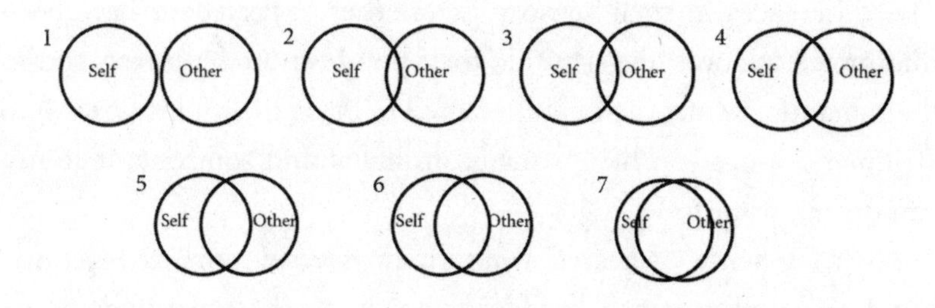

The higher their sense of shared reality with their chosen acquaintance, the more likely they were to choose the heavily overlapping circles to represent the relationship. Tellingly, the people who scored an average of 6 or 7 on the SR-G were also far more likely than lower scorers to respond positively to the explicit question: 'Have you ever felt that you and your partner had, in some sense, merged your minds?'

For further evidence, Rossignac-Milon attempted to disrupt people's sense of shared reality so that she could observe how they reacted. To do so, she invited couples into the lab and asked them to take part in a series of sensory tests – such as judging the smoothness of velvet or the sweetness and chewiness of a lychee gummy. After they had given their answers, the participants were told that a computer algorithm had crunched their data to provide a measure of the extent to which they overlapped in their experience of the sensory world. The feedback was a sham – deliberately engineered to suggest that the couples had very differing perceptions – but the participants fully believed what they were told.

As an outside observer, you might not expect a small difference in opinion over the sweetness of a gummy to be a serious cause for concern. Much like the Imaginiff questions, details like this don't seem to be very important in the grand scheme of things. Yet many participants – particularly those couples who had scored highly on the original SR-G test – appeared to be troubled by this tiny rupture in the shared reality they had constructed with their romantic partner, and they made active attempts to repair that feeling. They made more of an effort to validate each other's opinions during a subsequent conversation, for instance, and they started to talk about their common memories and in-jokes.

The differences in their sensory experiences appeared to have been interpreted as a warning sign that their bond was under threat, and so the romantic partners actively attempted to bring their minds back into alignment. Once you have a shared reality within someone, it seems, you do not want to let it go.

A strong sense of shared reality may grow over many meetings, but as Pinel also demonstrated, the seeds can be sown in minimal interactions. To test how we form a shared reality with strangers, Rossignac-Milon paired off more than 200 participants – previously unacquainted – and asked them to talk on an instant messaging platform. Their task was to discuss their interpretations of pictures depicting ambiguous events, such as a pencil drawing of two men talking at a bar. What were they talking about? And how would the scene unfold? After just twelve minutes of conversation, the participants were asked to leave the chat and to rate their feelings towards the other person, with questionnaires that included an adapted version of the SR-G scale.

Despite the very brief amount of time they had spent together, many of the participants had already started to establish the feelings of shared reality. The higher their score on the SR-G, the more likely they were to report having 'clicked' during the conversation, and the greater their reported desire to continue their discussion after the experiment had finished. The creation of the shared reality was also obvious in their outward behaviour. The pairs with the higher SR-G scores were more likely to say the same thing at the same time, to finish off each other's ideas in the conversation, and to converge on the same interpretation of the pictures. From the flow of the words in their dialogues, it looked as if they were both part of a single stream of consciousness.[13]

INTERBRAIN COUPLING

Rossignac-Milon's research on shared reality and the blurring of boundaries reminds me of Michel de Montaigne's essay 'On Friendship', a meditation on what it means to connect with another person and a beautiful elegy to the poet Étienne de La Boétie. The two men met by chance at a feast, and instantly formed a close bond that would last until the untimely death of La Boétie six years later, at the age of thirty-two.

Montaigne describes their relationship as an almost spiritual union. 'Not one of his actions could be set before me – no matter what it looked like – without my immediately discovering its motives,' he wrote. 'Our souls were yoked together in such unity, and contemplated each other with so ardent an affection, and with the same affection revealed each to each other right down to the very entrails, that not only did I know his mind as well as I knew my own but I would have entrusted myself to him with greater assurance than to myself.'[14] Elsewhere, he claimed that 'in the friendship I am talking about, souls are mingled and confounded in so universal a blending that they efface the seam which joins them together'.[15]

Montaigne is portraying an idealised friendship, but Rossignac-Milon's research shows us that a sense of unity in thought and feeling is central to many relationships. With advanced neuroimaging, we can even observe the merging of two minds in living brains. When people share a bond, they show remarkably similar neural responses to the same events – a kind of resonance known as 'interbrain coupling' that underlies their shared reality.

One of the best demonstrations of this process comes from Carolyn Parkinson at the University of California, Los Angeles and colleagues from Dartmouth College, New Hampshire. The researchers first recruited 279 graduate students from the same leadership course, and questioned them about their relationships with their classmates – who they were most likely to meet for lunch, drinks or visits to the cinema, for example.

From this data, the team was able to map out a social network and to estimate the nature of people's ties – whether they were close or distant. Around three months later, the researchers invited a subset of students into the lab to sit in an fMRI brain scanner while they watched a series of films – including an awe-inspiring sequence of an astronaut's view of Earth, a segment from a documentary on the food industry, a cringe-worthy clip of an inept comedian trying his hand at improv, and a heart-rending music video.

Given the constraints of the fMRI scanners, the researchers recorded each participant individually; they were not able to communicate with their friends or acquaintances as they watched the films. Even so, the researchers found that some pairs of participants showed remarkably similar neural responses to each clip, while others showed much less overlap. Two people's amygdalae might start to light up at the same scene, for example. This region of the brain is associated with emotional processing, and if two people were showing very similar patterns of activity in the amygdala, it would suggest they were interpreting the emotional content in a very similar way.

Parkinson found that the degree of overlap between the participants' neural activity could predict their relative positions in the social network. The more similar the two participants' brain responses were, the closer they were. As predicted by shared reality theory, close friends seem to have been paying attention to the same elements of the films and interpreting them in the same way – and this was reflected in the spontaneous fluctuations of their brainwaves. 'We are exceptionally similar to our friends in how we perceive and respond to the world around us,' Parkinson's team concluded.[16]

We can guess that if the participants had viewed the videos together at the same time, the neural resonance would have been even greater. One's laughter at a particular comedy could have prompted the other to find the film even funnier, or their tears could have heightened their friend's sadness – triggering even greater synchrony. Such experiments of real-time interactions are hard to achieve with the large and cumbersome fMRI scanners used by Parkinson's team, but researchers can use skullcaps that detect electric fields emanating from the underlying neural regions. Although this offers a relatively crude measure of brain activity, compared to the far more precise and detailed fMRI, the portability of the skullcaps allows researchers to witness the merging of minds in real time.

These experiments confirm that people with a greater sense of social connection show increased synchronisation in their brain activity during their interactions – whether we're talking about romantic couples, parents and children, or two strangers meeting for the first time. Much of the synchronisation occurs in the default mode network (DMN), which includes regions responsible for emotion, autobiographical memory, general knowledge and future planning. The DMN is thought to be involved in integrating outside information with our existing memories and schemas of the way the world works. That allows us to make sense of our current situation and decide how to respond appropriately. It should therefore be no surprise that similar reactions within the DMN will correspond to very similar interpretations of events.[17]

The role of brain synchronisation as the basis of a shared reality and social bonding has now been documented in many other contexts.[18] Interestingly, eye contact seems to trigger brain synchronisation – which may explain why it is so important for good communication. Somehow, when our eyes meet, we absorb the subtle emotional cues representing each other's mental states and our brains start to mirror each other.[19] In a very real sense, the eyes are 'windows of the soul'.

Besides helping to establish rapport and closeness, interbrain coupling can come with big advantages as we collaborate with other people. By ensuring that we pay attention to the same things and process them in the same way, the shared neural activity should make it much easier to coordinate our actions, rendering our exchanges smoother.[20] This is essential when working on a physical or creative task. It could mean that you can throw your partner the right tool without them having to name which one, or allow you to quickly riff off their idea without them having to explain their thinking ad nauseum.

The creation of a shared reality should also help to ensure that people provide the necessary emotional support. And that has important consequences for our wellbeing, with research showing the sense of being understood by others – and understanding them in return – underlies the link between social connection and health that we explored in Chapter 1.[21] The more we feel that we have a shared reality with someone, the better we function – and that is why the need to focus on the commonalities in our thoughts and feelings comprises our second law of connection.[22]

COLLECTIVE EFFERVESCENCE

The importance of shared reality for collaboration and cooperation may have been so great that we evolved specific behaviours to encourage its formation.[23] The origins of this theory can be found in the research of Émile Durkheim, the pioneering social scientist working in the late nineteenth and early twentieth century, who noted that many rituals involve large gatherings of people acting in synchrony. Collective chanting, drumming, singing and dancing are the most obvious examples. But even acts of passive spectatorship can trigger people's emotions to become aligned: just imagine the collective gasps and cheers of people watching a fire walker taking steps over hot coals. Durkheim proposed that these shared experiences would contribute to greater social cohesion, a phenomenon he described as 'collective effervescence'.[24]

Anything that aligns people's physiological and mental states should, in theory, encourage a feeling of shared reality that blurs the boundaries between the self and other, at least temporarily.[25] This can be as simple as synchronising our movements. Scott Wiltermuth and Chip Heath, who were both at Stanford University, put this idea to the test experimentally. In one study, they divided their participants into groups of three and asked them to take a turn around the campus. Some were allowed to walk naturally, while others were asked to ensure that their steps were in sync with each other at all times. On returning to the lab, the participants were then asked to rate how connected they felt to their new companions, and how much they trusted them, on a scale of 1 (not at all) to 7 (very much). Those ratings were considerably higher for the participants who had made sure that their steps beat the same rhythm on the pavement. The greater trust was also reflected in more cooperative behaviour during a laboratory game for small sums of money. After walking in synchrony, they were more likely to share prizes than to make selfish choices that might cause their companions to lose out.

As a second test of the idea, Wiltermuth and Heath asked their participants to sing along to the Canadian national anthem. (This may seem like an odd choice to inflict on a group of students living in the US, but it was deliberately chosen precisely because it was unlikely to have strongly favourable associations.) Once again, the shared action of singing led to feelings of greater affinity and more trusting behaviour.[26]

These effects have been replicated many times in multiple contexts among diverse groups of people. These studies have shown that even subtle rhythmic movements, such as arm curls, finger drumming or swinging in rocking chairs, as well as full-blown singing and dancing, are enough to increase measures of liking, compassion and cooperation between partners. When they were allowed to converse, the participants would often remember more information provided by the other person, and the increased affiliation could also be seen in their responses to the Inclusion-of-Other-in-Self scale that Rossignac-Milon used to chart feelings of shared reality. In many cases, just a few minutes of shared activity can lead to greater feelings of connection.[27]

Given these findings, many evolutionary psychologists argue that rhythmic song and dance evolved to enhance social bonding; they are part of the 'social glue' that held early groups of humans together and eventually enabled the extraordinary acts of cooperation and coordination necessary for the growth of society as we know it.[28] It's a beautiful idea that civilisation was built to the beat of a drum.

THE EXPANDING SELF

One of my favourite descriptions of shared reality comes from the singer Patti Smith. In her memoir *M Train*, she recalls the early days of her relationship with the guitarist Fred Smith, who died in 1994. She sometimes joked that she had only married Smith so that she would not have to change her name, and yet it is hard to imagine a stronger sense of connection as they wrote, built a home, travelled across the US and brought up their children. With each other's support, they achieved so much more than either would have accomplished individually. 'Looking back, long after his death, our way of living seems a miracle, one that could only be achieved by the silent synchronization of the jewels and gears of a common mind,' she writes.[29]

I love this description not only for its beautiful turn of phrase in describing Patti's shared reality with Fred, but also for the fact it captures the importance of 'self-expansion' in our most meaningful relationships: they can broaden and build our sense of who we are, so that we become a better version of ourselves. Husband-and-wife researchers Arthur and Elaine Aron have been at the forefront of the research on self-expansion. They point out that growing our personal abilities and resources is one of humans' basic motivations, and the most successful relationships – both platonic and romantic – help each individual to achieve this.

The Arons have now accumulated considerable evidence to support this hypothesis. They tracked hundreds of students over a semester, for instance, asking them to report their relationship status and to answer the open-ended question 'Who are you today?' using as many adjectives as possible. As you might expect of people in their late teens and early twenties, many of the students started falling in love over this period, and when that was the case, they started to use many more adjectives and nouns in their descriptions: their self-concept literally expanded, as their partners helped them to discover new aspects of themselves.

There are many ways that relationships – both platonic and romantic – could contribute to self-expansion. Once we have begun to build a sense of shared reality with someone, and established that our basic understanding of the world is the same, we start to integrate some of their characteristics into ourselves, and we may find that some of our initial differences become opportunities for learning and growth. When I, as a maths undergraduate, started dating an English literature student, for instance, my interest in poetry and fiction naturally increased through our conversations; when I became a friend of an environmentalist, I naturally started to care more about the natural world, because I saw how much it mattered to him.

Expansion of the self may also come from mutual encouragement to pursue our individual goals. We may not have been brave enough to follow our dreams before a friend or partner gave us the gentle push to at least flex our wings. And the pride that we feel in someone else's successes can contribute to our own sense of self-expansion. Finally, we may engage in new activities with our friends or partners that are self-expanding for both parties. This could be as simple as trying out new types of cuisine, attending art classes together or travelling to a distant country. Couples who regularly engage in new activities together tend to have greater relationship satisfaction and score more highly on the Inclusion-of-Other-in-Self scale, with a larger overlap between the circle representing the individual and the circle representing their partner. In these cases, the sense of growing together can actively contribute to the creation and maintenance of the couple's shared reality.

However it emerges, a feeling of self-expansion is one of the best predictors of closeness and satisfaction for all kinds of relationships.[30] We want our connections to engage in a mutual understanding *and* to be part of each other's growth.

CONSTRUCTING CONNECTION

This new understanding of shared reality and self-expansion offers some immediate suggestions for building better relationships. The first lesson is that we should be far more open-minded about the people we meet, since our usual predictions of who we will or will not like are often severely misguided.

Paul Eastwick at the University of California, Davis, for instance, asked participants to describe the three most important characteristics of their ideal partner. One week later, he then paired up the participants and placed them on speed dates. As you might expect, people often had very strong ideas about their ideal date. Whether or not someone met those criteria however, made very little difference to their romantic interest in the person. Indeed, Eastwick found that the participants were no more or less likely to feel chemistry with a person who had met their own criteria than someone who had met a stranger's criteria comprising three completely different traits. It was as if they had decided to eat out at a restaurant, ordered a specific dinner, and then swapped food with a person on the next table: they were equally likely to enjoy the random dish as the one they originally ordered.[31]

This pattern of behaviour is not restricted to dating or sexual attraction; our predictions are just as imprecise for platonic relationships.[32] We naturally look for people who fit certain profiles, and we have all kinds of presumptions about others based on factors like their social background, education and profession. But the creation of shared reality depends on specific thoughts and feelings; it's all about the idiosyncrasies of someone's mind and whether they overlap with our own, and that is very hard to predict upfront. To tell if someone shares our experience of the world, we have to give them a chance to reveal themselves to us, and we need to be willing to expose our own thoughts and feelings in return.[33]

In general, people who have more optimistic expectations about the possibility of forming a connection are more likely to overlook superficial differences, and wait to see if they feel the spark of connection. They tend to have more diverse relationships, while those with fixed assumptions tend to have more limited social circles.[34] Our beliefs about social connection can become self-fulfilling prophecies.

* * *

The second lesson from the research on shared reality concerns when, where and how we interact with others. We saw earlier how activities such as singing and dancing that promote a shared physiological or emotional response have long been used to promote bonding in large groups, and we can also use that to our personal advantage.

At the very least, this research offers an additional rationale for arranging an evening of karaoke with your colleagues, or a club night with your friends. (I find it telling that, in letters home during her first semester at Smith College, Plath reported the greatest sense of camaraderie with her housemates after an evening of singing around the piano: the joint activity chased away the existential isolation she felt on other evenings.)

People who live or work with young children might be interested to learn that singing and dancing can even encourage cooperation in the youngest minds. Fourteen-month-old babies, for instance, were more helpful to an adult experimenter after they bounced together in time to an instrumental version of the Beatles' 'Twist and Shout'. They were more likely to pass a marker pen that the scientist needed, or to pick up a clothes peg that the researcher had 'accidentally' dropped on the floor.[35]

The benefits of behavioural synchrony can even be incorporated into the workplace. One German publisher recently instigated a nine-week exercise programme for employees that gently encouraged teams to engage in co-ordinated movements. Questionnaires revealed that the employees taking part in the regime felt closer to their colleagues, and they experienced less of the usual irritation that inevitably arises from office life. Perhaps because of their reduced stress, they also took fewer sick days.[36]

If singing and dancing in public is not your style, don't despair, many other activities will help encourage the physical and emotional resonance that sparks a basic sense of social connection. Anything that focuses your attention on the same thing at the same time as another person, and provokes the same visceral reaction, will help to strengthen bonds. Comedy, for instance, produces synchronised laughter, which can also offer a powerful means of establishing a shared reality – provided you can find something that appeals to both of your senses of humour to ensure that you do appreciate the same jokes. (Sitting next to someone in stony silence, while they guffaw at the routine, is not likely to create a bond.) The synchronisation of emotional highs and lows can also explain why sports fans find events to be such strong bonding experiences. If you support the same team, the tension and exhilaration will realign your internal state so that you feel much closer at the end of the event.[37]

Even physically painful experiences may do the job. Certain initiation rituals may work like this: when we know that the people around us are experiencing exactly what we are feeling at that moment, agony *is* the shared reality, and our recognition of that fact increases our liking and trust for the other person. It doesn't have to be extreme. For people who are relatively unaccustomed to spicy food, the eating of uncomfortably hot dishes can serve this purpose. One study found that strangers who ate chilli peppers together tended to show greater social connection than those who tasted candies.[38] This might explain why certain dishes, such as Sichuanese hotpot – famous in China and increasingly popular in the West – are often used as a bonding experience. It is not just the undoubtedly delicious and complex flavours of the food, but also the joint discomfort of the fiery spice, that temporarily unites us.

Another route is to look for activities that bring a sense of growth and self-expansion. It could be stargazing or life drawing or competing in ultramarathons; the choice will be entirely dependent on your tastes, but ideally you should look for something unfamiliar that neither of you has tried before. Multiple studies have shown that people who undertake challenging activities together tend to feel closer and more committed to each other, compared to those who spend their time together more passively.[39]

Whichever strategy you embark on, you should pay close attention to your partner's responses and – if you share them – affirm that you think and feel the same way. You may be surprised by just how similar your experiences are, and that validation will enhance your feelings of closeness.

In a journal entry from 1937, the writer Anaïs Nin mused on her relationship with Gonzalo Moré, a Peruvian activist who awakened her political sensibilities. 'With Gonzalo I rediscovered my Spanish world, my Spanish blood, warmth and personal involvement, direct passionate response to experience, fire, fanaticism, fervour, faith, the power to act whole, wholeness, caring,' she wrote. 'How each friend represents a world in us, a world possibly not born until they arrive, and it is only by this meeting that a new world is born.'[40]

Psychologists refer to these worlds as the people's shared reality – and as Nin suggests, we can occupy different shared realities with different people, each one bringing out different qualities within us. The writings of Sylvia Plath, Michel de Montaigne and Patti Smith all help to illustrate how the presence or absence of shared reality with another person contributes to connection or alienation.

We have already identified a few different strategies for building a shared reality, but as a governing principle, our second law of connection can be summarised like this: **Create a mutual understanding with the people you meet. Ignore superficial similarities and instead focus on your internal worlds, and the peculiar ways that your thoughts and feelings coincide.** Over the following chapters, we'll explore further ways to maintain and grow a shared reality through our conversations, and we'll learn how to heal the small fractures of minor disagreements and the larger ruptures that can be created by major disagreements. We'll start in Chapter 3 by looking at the 'myth of personality'. Due to shyness, you may feel that you simply can't break through the barriers that are necessary to reach a shared understanding. This is an extremely common fear – but it's also a myth that blinds us to opportunities for connection. Life is not loneliness, as the teenage Plath claimed; it is full of rich possibilities for developing meaningful relationships, provided we know how to look for them.

What you need to know

- Shared reality is the feeling that someone experiences the world in the same way that you do. Even if you come from very different backgrounds, your thoughts and feelings are somehow aligned and occurring in tandem
- The feelings of shared reality between two people can predict their immediate rapport and the long-term closeness and commitment to their relationship
- Shared reality is reflected in 'interbrain coupling', in which two people's neural activity starts to resonate with the same responses to the same events
- Relationships are strongest when shared reality is combined with a feeling of self-expansion, so that each person is helping the other to grow

Action points

- Assess your shared reality using the SR-G scale on pages 35–6 with a few of your acquaintances. This may help you to identify situations in which you could make more of an effort to build rapport
- Consider whether you do your best to validate the thoughts and feelings of the people around you. For example, do you make it obvious when you agree with someone's point of view, or do you tend to stay silent? These tiny actions may seem insignificant, but they emphasise the fact that your brains are working in synchrony
- If you wish to build more rapport with someone, consider undertaking activities that contribute to the creation of shared reality. These are activities such as music, dance or comedy that synchronise your emotional and physiological reactions through movement or laughter
- Consider embarking on a new and fresh challenge with someone you love to boost your sense of self-expansion

CHAPTER 3

THE PERSONALITY MYTH

Jane Austen's Mr Darcy is known for being exceptionally handsome, proud and aloof, but does his behaviour represent a common barrier to social connection? Around halfway through *Pride and Prejudice*, Darcy and Elizabeth Bennet have a telling conversation about the nature of social skills. Elizabeth is sitting at the piano, and as she plays, she teases Darcy for remaining reserved and cold at social gatherings. He does not deny the accusation but declares that he is simply 'ill-qualified to recommend himself to strangers'. 'I certainly have not the talent which some people possess of conversing easily with those I have never seen before,' he says. 'I cannot catch their tone of conversation, or appear interested in their concerns, as I often see done.' Elizabeth's response is scornful. 'My fingers do not move over this instrument in the masterly manner which I see so many women's do. They have not the same force or rapidity, and do not produce the same expression. But then I have always supposed it to be my own fault – because I would not take the trouble of practising. It is not that I do not believe my fingers as capable as any other woman's of superior execution.'

Multiple strands of research show that Darcy's opinions are widely held by many people today. They assume that they are not socially competent enough to talk to new people and that the conversation will be awkward and embarrassing for all concerned. As a result, they think it would be better not to make the effort.

In this chapter, I want to show you why Elizabeth is correct: much like playing a musical instrument, making new acquaintances becomes easier with practice. And even if your first attempts are a little clumsy, the effort pays off. It is more pleasurable to create a shared reality with someone than you might think, and you will feel better connected as a result, with all the consequent benefits for your health and wellbeing. Even if you never see that person again, the exercise will have boosted your confidence in your social skills.

FIRST IMPRESSIONS

Let's begin with a philosophical experiment that shows just how keen people are to form a connection and to smooth over slight off-notes in social interactions. The study takes inspiration from Edmond Rostand's play *Cyrano de Bergerac*, first performed at the end of the nineteenth century. By the end of Act II, the handsome-but-dim Christian enlists the help of his friend, the poet Cyrano, to help woo the beautiful Roxane. Cyrano is also secretly in love with Roxane, but he feels he is too ugly to have a chance, and so he agrees to help Christian by ghostwriting witty repartee that will help win her heart. In the play's most famous scene, Roxane is standing on her balcony while Christian expresses his devotion from below. Little does Roxane know that Cyrano is whispering the words in Christian's ear.

In the mid-2010s, Kevin Corti and Alex Gillespie at the London School of Economics decided to replicate the basic scenario of Rostand's play, but in place of the witty poet Cyrano, they used an artificially intelligent chat-bot, which fed lines to an actor through a headpiece. The actor then chatted with experimental participants, who had no idea of the experimental set-up, and who were subsequently questioned about their interactions.

Artificial intelligence has made great leaps over the past decade – but in the mid-2010s it was still very clunky, meaning that the conversations were filled with non-sequiturs and strange phrasing. Astonishingly, however, most participants made a substantial effort to establish common ground and repair the conversations when they took an unexpected turn.[1]

I first wrote about these 'echoborg' experiments in 2015, and during my research for that article I even fell for the trick myself. On entering the lab, I was greeted by a student, Sophia Ben-Achour, who made slightly stilted small talk before Gillespie and Corti entered the room. It was only then that I realised that a chat-bot had been talking through Sophia. Like many of the participants in their study, I'd been fooled, and I had been willing to make all kinds of allowances for some of the odd turns our conversation had taken. I found this strangely reassuring for my own social anxieties. Surely if people could be so forgiving of a clumsy chat-bot – and mistake it for a real human mind – they would also be willing to overlook my occasional faux pas?

Little did I know that parallel research from other social psychology labs was beginning to show exactly this. In one eye-catching study, Nicholas Epley and Juliana Schroeder at the University of Chicago recruited around 100 commuters taking a train from the local village of Homewood, and asked them to follow one of three instructions. Those in the 'connection condition' were given the following guidelines: 'Please have a conversation with a new person on the train today. Try to make a connection. Find out something interesting about him or her and tell them something about you. The longer the conversation, the better. Your goal is to try to get to know your community neighbour this morning.' The second group were told to 'keep to yourself and enjoy your solitude . . . Take this time to sit alone with your thoughts. Your goal is to focus on yourself and the day ahead of you.' The third group were simply asked to act as they normally would on a normal day – a comparison that allowed Epley and Schroeder to control for any changes that might come from simply breaking one's normal routine to try something new and different.

After they had completed the commute, the participants were asked to return a questionnaire about their immediate thoughts and feelings. Not one of those in the connection condition reported having had a negative interaction, suggesting that truly hostile reactions are extremely rare. The conversations were, in fact, uniformly pleasant, meaning that they left the journey feeling far more positive than those who had been asked to enjoy their solitude, or those who had simply carried out their commute as normal without any marked change to their behaviour. These experiences were a stark contrast to participants' predictions of what the different commutes would be like; when asked to imagine the different scenarios, most people expected to relish a bit of extra 'me-time' without others' company adding to the day's stresses. To confirm the result, Epley and Schroeder repeated the experiment with participants taking a bus journey to their lab. They found the same responses: beforehand, people were pessimistic about the chances of connecting with someone they didn't know, yet the experience proved to be far more enjoyable than they had anticipated.[2]

After Epley and Schroeder published their findings, some critics questioned whether people in the UK would be similarly receptive to approaches. Londoners, in particular, have a reputation for reserve and unfriendliness – particularly in the cramped conditions of the Tube. When the same researchers encouraged British passengers to make a connection, however, the conversations once again proved to be more enjoyable than expected. The alleged cultural differences between the US and UK did not diminish the effect. One particular barrier was perceived interest: the participants doubted that the other commuters would want to chat and thought that they'd be considered a nuisance for trying to strike up a conversation, but this was rarely the case.[3]

The phenomenon is by no means unique to the peculiar setting of public transport. While working at the University of British Columbia, Gillian Sandstrom and Elizabeth Dunn stood in front of a Starbucks coffee shop and asked customers to exchange a few friendly words with the barista – a small interaction that temporarily buoyed the customers' mood compared to those who simply gave their order in the most efficient way possible.[4]

Sandstrom has since replicated the findings in a range of experiments involving more than 2,300 people in total. The volunteers included students in the lab, people on a personal development course and members of the general public. The interventions ranged from single conversations to week-long quests to make more connections with multiple strangers. In each case, people's fears about making a connection were found to be 'vastly overblown' and the conversations – either short or long – proved to be more rewarding than anticipated.[5]

THE LIKING GAP

If we are to fulfil our social potential, we must also extinguish the anxieties that come *after* our interactions. Following a perfectly pleasant conversation, many of us find that a dark cloud of doubt continues to overshadow the afterglow of a social interaction. No matter how enjoyable we found the interaction, we can't help but worry about the way we were perceived by the other person. Even if we managed to build a shared reality, our doubts can erode the sense of mutual understanding, discouraging us from building a deeper connection.

This is the 'liking gap' that I mentioned in the introduction. As you may remember, the concept of the liking gap comes from Erica Boothby and Gus Cooney, who asked people to converse in various contexts and then questioned them about how much they liked each other. They found that the average person leaves a conversation feeling that their new acquaintance liked them far less than they liked their new acquaintance – and vice versa. This is a consistent gulf that means we are overly pessimistic about the impressions that we have made, and the chances of forming a friendship in the future.

The liking gap would not be such a serious problem if it disappeared on a second meeting, but these doubts about the strength of a social connection can linger weeks or even months into a new acquaintanceship. One of Boothby and Cooney's studies, for example, examined the impressions of 102 roommates at Yale University, with questionnaires issued in September – when they first met – and follow-ups in October, December, February and May. They were asked to rate on a scale of 1 (not at all) to 7 (very much):

- How much do you like your suitemate?
- How interested are you in getting to know them better?
- How interested are you in becoming better friends with them?
- How interested are you in spending more time with them?

They next answered a set of questions on the same themes, but with the roles reversed, so they had to guess how much their suitemate liked them, wished to spend more time with them, wanted to know them better and hoped to become friends.

At the start of their acquaintance, most people underestimated the warmth of their suitemate's responses to all these questions – replicating the liking gap. And those doubts faded extremely slowly. Indeed, it was only by May – eight months after their first meeting – that they had stopped underestimating how much their suitemate appreciated them.[6]

The liking gap can be a source of disconnection and demotivation in the workplace. Boothby and Cooney recruited teams of engineers and questioned each individual about their relationships with the other members of their group. They found that the liking gap was prevalent among these teammates, even if they had worked closely together on a project for a few months. And this seemed to be a barrier to effective collaborations. The greater the liking gap, the less likely people were to ask their teammate for help or to give them honest feedback. It also limited how much they wanted to work together again. Further surveys, of a more diverse group of subjects, found that the liking gap also contributed to reduced job satisfaction.

In written descriptions of their relationships with their colleagues, it was striking how easily the participants could name the positive qualities of the people around them while doubting their own attributes. They'd praise a workmate for being 'straightforward', for example, while damning their own attention to detail. 'He thinks I can be annoying because I am way too concerned about performing well and accurately,' one wrote. This participant didn't seem to consider the possibility that their work ethic could also be seen as a positive attribute to their colleague.[7]

The liking gap seems to be common for people of all genders, and it emerges at a young age. Wouter Wolf, from Utrecht University in the Netherlands, has found that children aged four do not tend to worry about they way they will be perceived by others. By the age of five, however, they have started to absorb the concept of politeness, and they become increasingly aware that other people may be hiding how they feel; when someone seems excited by what they say, the child begins to understand they may just be feigning interest. With this realisation, they start second-guessing others' responses; they begin to worry that any feeling of mutual understanding and appreciation is simply an illusion.[8]

When interacting with others, it is inevitable that we will occasionally say something that is *genuinely* clumsy. Boothby and Cooney's research on the liking gap doesn't explicitly explore these eventualities, but a simple thought experiment can help us to see that our worries about objective faux pas are also exaggerated. Imagine, for instance, that you are a guest at a dinner party, and you are the only person to turn up without a gift for the host. How negatively do you think you would be judged on a scale of 0 (not at all) to 10 (a great deal)? Now put yourself in the shoes of the host. How negatively would you judge a person who turned up to your party empty-handed, on the same scale of 0 to 10?

You probably assumed that other people would be much harder on you than you'd ever be on your own guests. Experimental participants who were asked to consider this scenario tended to believe that others' ratings of the faux pas would be about twice as negative as their own would be of someone making the same mistake – with average ratings of 5.26 compared to 2.47 on that 10-point scale. That might not sound so striking until you frame the result more positively: on average, others' judgements of our mistakes will be *half* as severe as we assume. Happily, we see the same misaligned predictions in many other scenarios. Whether we are setting off a public alarm in a shop or library, revealing our ignorance on a general knowledge quiz or being seen in an embarrassing piece of clothing, people are not nearly as judgemental as we expect.[9]

Many of our social anxieties can revolve around our responses to the embarrassment, rather than the embarrassment itself. I easily blush, for instance – and in the past, I would become fixated on the heat rising in my cheeks; the reddening face, I assumed, was only drawing attention to a mishap and making me look even more foolish. Research shows that these kinds of fears are common, and once again, they are unfounded. While blushing can betray a lack of self-confidence, that doesn't reduce people's likeability, and in some situations may even increase it. Researchers have asked participants to read about various types of mishaps that might leave someone red-faced – such as knocking over a rack full of wine glasses in a shop. When the participants are told that the person blushed deeply, or are shown a photo of someone with a flushed face, they tend to rate the person as being more friendly and more reliable. Seeing someone blush also increases the perceived sincerity of their apology, and leads onlookers to consider the transgression itself to be less serious.[10]

Many of the other overt signs of social nerves can be similarly appealing. When you feel that you're the centre of attention and might be judged negatively by others, you might have the habit of touching your face, running your fingers through your hair, licking your lips or fiddling with your wedding ring. While you may wish to suppress anxious feelings and present a more confident front, these behaviours are generally greeted with sympathy rather than scorn. In one study, participants were asked to take part in an ordeal known as the 'Trier Social Stress Test', which involves giving a presentation, taking part in a fake job interview and performing on-the-spot mental arithmetic. The participants with the most obvious signs of nerves were considered the most likeable by a group of independent judges, compared to those with cooler veneers.[11]

MISPLACED 'META-PERCEPTIONS'

If you are familiar with the psychological research on cognitive bias, you'll find these discoveries to be a striking contrast to the abundant literature on over-confidence. When asked to rate their own abilities compared to others', people tend to be unrealistically optimistic. The 'better-than-average effect' has now been documented in people's ratings of everything from intelligence to driving skills and even morality; we certainly aren't suffering from generally low self-esteem.[12] That may seem difficult to reconcile with our under-confidence in social situations, but careful analysis suggests that it depends on perspective: how we view ourselves versus how we think others view us. Put simply, we see ourselves with rose-tinted spectacles, but assume that others view us through darkened shades that filter out all our good qualities. As a result, we can feel privately confident but we also fear situations in which we might be judged by critical eyes.

Multiple scientific papers have now demonstrated that the liking gap is just one example of a far more general problem with estimating how others see us. Scientists describe these collectively as 'meta-perceptions', since they are perceptions about someone else's perceptions. One Canadian study placed more than 2,000 participants into pairs or groups and then measured how accurately they could judge the impressions that they made on each other. Each person had to rate how the other participants would have assessed their intelligence, for example, or their humour. For almost every trait considered, the participants showed a 'negativity bias'. They assumed that they came across as less smart, less funny, less conscientious, less open-minded and less agreeable than others considered them to be.[13] When imagining how they looked through others' eyes, they saw someone rather dismal.

Regarding social skills, our negative meta-perceptions may be driven by our consideration of two different characteristics: *competence* – how eloquent and entertaining we are, for instance – and *warmth* – whether we seem kind and generous. Like Christian in *Cyrano de Bergerac*, we may overemphasise the importance of competence. We listen to our inner critic – that internal voice that expects us to act with perfect social grace in all situations – and assume that we'll be judged harshly for any blunder. We forget that people will also be considering how warmly we appear to them: whether we validate their thoughts and feelings and show concern for their needs.[14] 'People may forget what you said, but they will never forget how you made them feel,' as the common saying goes.[15] And once the conversation is over, we blame any moment of awkwardness on ourselves, while forgetting that the other person may also be doing the same.

It is easy to see how these biased meta-perceptions could all pose a serious barrier to connection. A student starting university, for example, might strike up a conversation with the person sitting next to them in a lecture theatre, but worrying that their new acquaintance had only feigned interest, they would fail to invite them to continue the conversation after class. At work, our concerns about appearing unintelligent or incompetent could prevent us from reaching out to colleagues on another team. Or, after speaking out of turn at a party, we might read so much into that tiny gaffe that we deliberately avoid seeing our interlocutor again, when they had quite forgotten our clumsy mistake. In each case, we're missing vital opportunities to expand our social circle, with all the benefits that would bring.

There will, of course, be times when other people don't warm to us as much as we'd wish. It would be foolish to think that we will connect with everyone we meet, and it should go without saying that we must be respectful of others' boundaries and be mindful of our potential to offend. The existence of the liking gap, and the knowledge that other people are generally forgiving of our social mishaps, does not mean that we can act insensitively without repercussions.

But the science shows we needn't be as pessimistic as we currently are. We are more liked and respected than we think, and by daring to recalibrate our expectations even a small amount, we may be pleasantly surprised by how quickly we can close the liking gap to build new bonds.

THE PERSONALITY MYTH

I hope that this knowledge of the liking gap may have already increased your confidence – it certainly boosted mine. If the thought of meeting new people still fills you with dread, take heart from the fact that it will become easier with time.

In one week-long challenge, Gillian Sandstrom's participants downloaded an app that set out various 'scavenger hunt' missions. On different days they were encouraged to find someone with interesting shoes or eye-catching hair, for example, and then talk to them for a few minutes. Sandstrom found that, day by day, the participants enjoyed the experiences and updated their beliefs accordingly. The more they spoke to strangers, the less anxious they felt about the possibility of rejection, and the more they came to discover their capacity to build a connection with people they didn't know.[16]

What's true in general might not be true for the individual, of course, and the Mr Darcys of this world might suspect that they are the exception that proves the rule. If you've always been shy and retiring, it's natural to question whether you have what it takes to make new connections. Until recently the scientific consensus seemed to provide some support for the idea that we cannot choose to become more gregarious. Psychologists have long included sociability as one of the characteristics of extroversion, which is one of the so-called Big Five personality traits that are thought to dominate our behaviour. People with low extroversion – commonly known as introverts – tend to be more reserved and more inhibited, while those who score highly on the trait tend to be chattier and more assertive and find it easier to make friends.

Thanks to their wider social circles, extroverts enjoy a boost to their overall wellbeing.[17] In the past, personalities were thought to have set firm in childhood and to be resistant to manipulation, meaning that introverts hoping to glean those benefits would struggle to go against their grain and act more sociably. Like the proverbial leopard that cannot change its spots, dyed-in-the-wool introverts would do better to accept their nature than attempt to expand their social behaviour.

The latest research, however, shows that introverts and extroverts can both benefit from exercising their social muscles more regularly.[18] Consider a study by Seth Margolis and Sonja Lyubomirsky at the University of California, Riverside which confirmed the conclusion in a study of 131 undergraduates. The study took the form of a two-week challenge. For the first seven days, half were told to 'act as talkative, assertive, and spontaneous as you can' (the behaviours of a classic extrovert) while the rest were told to be 'deliberate, quiet and reserved' (the behaviours of a classic introvert). The participants then switched – so that those who had tried to behave extrovertedly started acting more introvertedly, and vice versa. The effects of the intervention were considerable. Whatever their initial level of introversion or extroversion, the participants' positive mood and sense of connectedness increased when they acted more gregariously.[19]

Further research suggests that the big difference between introverts and extroverts does not concern the effects of social activity themselves, but their *expectations* of what they will feel. Before a social event like a cocktail party, introverts tend to have the poorest predictions about how much they will enjoy the interactions; unlike the extroverts, they assume that they'll feel considerably worse than they did before the event. Afterwards, however, the vast majority of introverts find the experience to be more fun and energising than they had anticipated. Introverts also expect that social events will deplete their cognitive resources, leaving them less able to focus, but psychological tests find no evidence for this belief, either.[20]

As Susan Cain so beautifully documented in her book *Quiet*, there are many advantages to introversion, and I'm certainly not claiming that anyone should feel obliged to become more boisterous or dominating if they are happy with their life the way it is. It would be counterproductive to behave in a way that does not feel authentic to your core self simply because you feel compelled to meet others' expectations. This research simply shows that change is within your power, if you want it, and that it can be rewarding for those who currently feel frustrated with their capacity to connect.

Whatever your personality, you can be selective about the situations in which you test out your social confidence. You may not choose to chat with every stranger in the park, but you might make a bit more effort talking to your local barista as they make your coffee, or with your hairdresser during your monthly trim. You might agree to attend a party or work gathering that you would have once made great efforts to avoid. Knowing that your 'affective forecasting' is often off-kilter should provide the reassurance that you will find these experiences more enjoyable than you expect, and that you will be liked more than you think.

If you are serious about changing your habits, it's worth formalising these steps with concrete 'implementation intentions' – a well-accepted psychological strategy for bringing about personal transformation. These come as 'if-then' statements, with a specific trigger and a concrete plan of action. For instance, you may decide: 'If I see someone in the park with a dog, I will try to talk to that person about their pet.' Or 'If I notice someone looking lost outside the station, I will offer to give them directions.' Generate as many of these as you can and then aim to perform a certain number a week. The research suggests that the more you put it into practice, the greater the benefits.[21]

Once you have started to grasp these extra opportunities to connect, you may become so used to your new-found sociability that you forget that you were ever nervous about new interactions. Alongside their studies of Chicago's train and bus users, Epley and Schroeder also spoke to taxi passengers about conversations during their journeys. They found that people who regularly made the effort to engage with the drivers were much more optimistic about how those encounters would play out, compared to those who had not yet developed this habit.

GROWING SOCIAL GRACE

Based on all these findings, our third law of connection can be summarised as follows: **Trust that others, on average, will like you as much as you like them, and be prepared to practise your social skills to build your social confidence.**

However you plan to apply this principle, recent psychological research offers a few strategies that should render the process more manageable. Your first move should be to curb the self-criticism that can blind us to the potential opportunities in front of us. You can do this using a strategy called 'cognitive restructuring', which involves questioning your assumptions and the ways you frame events.[22] Before attending a party with people you don't know very well, for example, you might find yourself thinking about all the ways the other guests might judge you, and you might be telling yourself 'no one likes me' or 'I'm going to make a fool of myself'. To break that thought cycle, you might try to remember a similar situation in which your expectations were pleasantly confounded. If you notice that someone is looking at you, you might start to feel self-conscious and suspect that there is something wrong with your appearance. But you could try to remind yourself that there are many other explanations. Perhaps you simply look familiar, and they are trying to place you, for instance.

As we have seen, many of us become far too fixated on the slightest mishap or embarrassment, but a little effort to consider the bigger picture reduces this tendency. This can be as simple as quickly listing all the different factors that can influence a first impression, including those that are beyond your control – such as the other person's mood, whether they have had enough sleep and whether you remind them of someone they already know. The more factors you recognise, the more obvious it becomes that a single clumsy remark is unlikely to dominate their opinions – and the less anxiety you feel. [23]

You should be particularly mindful of over-general, catastrophising thoughts. To overcome this, you can try to give yourself a 'reality check'. In this case, you might consider the very worst scenario that you fear, and then ask the following questions:

- How bad would that really be if it happened?
- What are the odds of that happening?
- How would I cope if the worst came to pass?

This should help you to look at the consequences a little more objectively so that you recognise your capacity to deal with the situation, whatever it throws at you.[24]

Along the way, you might try to cultivate a little self-compassion in your life. If you currently baulk at the concept, you are not alone. As the American-British comedian Ruby Wax wrote in her book *Frazzled*: 'When I hear of people being kind to themselves, I picture the types who light scented candles in their bathrooms and sink into a tub of Himalayan foetal yak milk.'[25] In reality, practising self-compassion simply means that you respond to your own failures and feelings of inadequacy with warmth and understanding, without resorting to self-flagellation or sweeping judgements about your overall value as a human being. It also involves recognising the 'shared humanity' of a situation: the fact that many people will feel the same way as you.

Many people mistake this for self-indulgence and believe that self-criticism is essential to learn from our mistakes. Yet the very opposite is true. People who are high on self-compassion, and low on self-criticism, are in fact *more* likely to change their behaviours. Students with high self-compassion tend to work harder after failing a test, for instance, and after having offended someone, people with high self-compassion are more likely to make amends for their mistakes. This may be because self-compassion allows us to think more proactively about the stressful event without being overwhelmed by anger or shame – demotivating emotions that encourage procrastination and avoidance.[26]

It's easy to see how this could be relevant to your social interactions. Making new connections can leave you feeling a little vulnerable, and someone high in self-criticism will be far more scared of failure and will judge themselves more harshly if they do not get the response they desire. As a result, they may believe that it's best to avoid the perceived risk of rejection; rather than sticking their neck out, they'll hide within their shell.[27] People with higher self-compassion, in contrast, tend to show reduced anxiety when they are conscious of being judged by others – and this is apparent in their levels of the stress hormone cortisol.[28]

You can tell if you have low or high self-compassion by considering the following statements, which are part of a twenty-six-item test.[29] In the scientific research, they would be rated on a scale of 1 (almost never) to 5 (almost always).

- I try to be loving toward myself when I'm feeling emotional pain
- I try to see my failings as part of the human condition
- When something painful happens, I try to take a balanced view of the situation

and

- I'm disapproving and judgmental about my own flaws and inadequacies
- When I think about my inadequacies it tends to make me feel more separate and cut off from the rest of the world
- When I'm feeling down, I tend to obsess and fixate on everything that's wrong

The more you agree with the first set of statements, the higher your self-compassion, and the more you agree with the second set, the higher your self-criticism. The good news for the lower scorers is that self-compassion can be trained. Try to spot when your inner voice is needlessly negative, and then consider what a friend or family member might say in its place. During a moment of embarrassment, for example, you can avoid making sweeping judgements about your social skills, and instead remind yourself that almost everyone feels awkward occasionally. Embarrassment is a universal experience and doesn't make you any less valued as a person.

To reinforce this new way of thinking, you might compose a short letter to yourself in which you deliberately adopt a kind and non-judgemental attitude about a situation that is upsetting you. Studies show that these writing exercises can bring about a change in attitude and effectively reduce people's anxieties about social connection.[30]

Last, but not least, you should consider your mindset – that is, your beliefs about personal development, which can shape how our brains respond to new challenges. This research originates with Carol Dweck at Stanford University, who initially focused on people's assumptions about intelligence and academic achievement. Some people, she found, believed that their skills were innate and unchangeable; they had a so-called 'fixed mindset'. Others saw their abilities as something that flourished over time, with practise; they were said to have the 'growth mindset'. Dweck and her colleagues found that people with a growth mindset have greater persistence when they face difficulties in their studies and are more willing to take on new opportunities, even if they push them out of their comfort zone. Those with fixed mindsets tend to find it harder to cope with setbacks and are more easily discouraged.

It's now becoming increasingly apparent that fixed and growth mindsets can influence many other important outcomes besides academic success.[31] Having a growth mindset can leave people feeling empowered about their capacity to cope with issues such as anxiety and depression, for example, meaning that they respond better to positive interventions, and the same may be true for people's fears about their social competence. Psychologists at Northern Illinois University, for instance, developed a scale examining people's mindsets about their shyness. To get a taste of this research, consider the following statements, which participants were asked to rate on a scale of 1 (strongly disagree) to 5 (strongly agree):

- You have a certain amount of social grace and you can't really do much to change it
- You can learn to get along with people better, but you can't really change how much people will like you

As you might have guessed, higher ratings reflect a fixed mindset about your shyness, while lower ratings reflect a growth mindset.[32] People with a fixed mindset tend to respond less well to interventions that encourage greater sociability, while those with a growth mindset see greater benefits.[33]

Fortunately, knowledge is power. When people are taught about the brain's capacity for change, and the ways that social skills can improve with practice, they are more likely to develop a growth mindset with all its benefits.[34] The cutting-edge research we have explored in the past few pages should have already primed you with a more positive mindset about your potential to change your social attitudes and behaviours – and if you take that knowledge with you as you venture outside your comfort zone, and regularly remind yourself of your capacity to grow, you may be surprised by what you achieve. As Elizabeth Bennet argued to Mr Darcy in *Pride and Prejudice* two centuries ago, our capacity to connect is no more fixed than our musical ability.

What you need to know

- People's fears about talking to strangers are most often unfounded. On average, you'll find these interactions to be far more rewarding than you expect
- After social interactions, we underestimate how much people like us

– doubts that can erode the sense of shared reality and discourage the formation of a potential friendship or a professional collaboration

- We are also unduly pessimistic about how people assess our intelligence and sense of humour. On average, people respect you more than you think
- Introverts tend to have lower expectations of social interactions than extroverts. After such meetings, however, they report just as much enjoyment
- People of all personality types can build greater social confidence

Action points

- Choose a situation in which you would prefer to be more outgoing and set out specific goals for your new behaviour. You may decide to make conversation with someone new in your work canteen, or to talk to every dog owner you see in the park
- Whenever you find that negative expectations are acting as a barrier to social connection, practice the cognitive restructuring exercise on page 64. Try to remember all the multiple factors that can determine someone's first impression. This will take your focus away from yourself, reducing your self-consciousness
- If you feel socially anxious, try to build more self-compassion in your life. For example, you may attempt to write a compassionate letter to yourself, framed as if you are giving advice to a friend. You can find a list of other resources in the appendix to this book

CHAPTER 4

OVERCOMING EGOCENTRIC THINKING

If J. Edgar Hoover's staff knew one thing about their boss, it was his impatience: he had little time for debate or discussion. Hoover had overseen the transformation of the FBI from a bureaucratic backwater into the world's most respected – and feared – intelligence and security service. He ruled with an iron fist and his judgement was considered the final word on every matter.

And so, when he returned a memo on internal security with the words 'Watch the borders' scrawled in his handwriting, his agents knew to take the message seriously. They just didn't know how, and given his reputation, they knew better than to ask for clarification. Phones soon started ringing across the building as everyone asked what was happening in Mexico or Canada, yet no one had observed any unusual activity. They contacted Customs and the Immigration and Naturalization Service but were still none the wiser.

According to his assistant, Cartha DeLoach, it would take days for his staff to get to the bottom of the mystery, when a supervisor noticed that Hoover had scrawled his message on a memo with very narrow margins. The director had a habit of writing his comments in the white space around the edges of the page. To Hoover, the meaning could not have been clearer: the command 'Watch the borders' was drawing attention to the page's sloppy formatting. Everyone else, however, assumed he had been talking about the borders of the country, and an imminent geopolitical crisis.[1]

The anecdote may tell us a lot about the perils of Hoover's management style, but it is even more illustrative of the communicative barriers that we all face.[2] All too often, we are trapped in our own way of thinking and misjudge how others will interpret our words or actions. This tendency is known as egocentric thinking, and it can lead us to wildly different conclusions about the same events. The effects of egocentric thinking can sometimes seem farcical: comedies of errors are full of characters operating at cross-purposes, with hilarious scenes resulting from their one-sided views of events. But the ramifications for our relationships are often serious. If left unchecked, egocentric thinking erodes the sense of shared reality and contributes to needless disagreement.

We've just learned how most of us are much more adept at making friends and strengthening our relationships than we might expect. But if we are to make the most of that social potential, we must recognise egocentric processing in its various manifestations and arm ourselves against it. This will be the basis for our fourth law of connection. With just a little practise, we can vastly reduce the risk of misunderstanding, resulting in the smoother and more meaningful interactions that bolster the close social connection we crave.

ARE YOU THINKING WHAT I'M THINKING?

To understand the origin of our egocentric mistakes, we must first recognise the mental processes that are typically required to understand another's point of view, and the ways that they can go wrong.

Since the 1970s, psychologists have described our capacity to recognise others' beliefs and perspectives as a 'theory of mind'. For most neurotypical people, it is thought to emerge in early childhood. Psychologists often measure it with the famous Sally–Anne test, presented through the following comic strip.

1

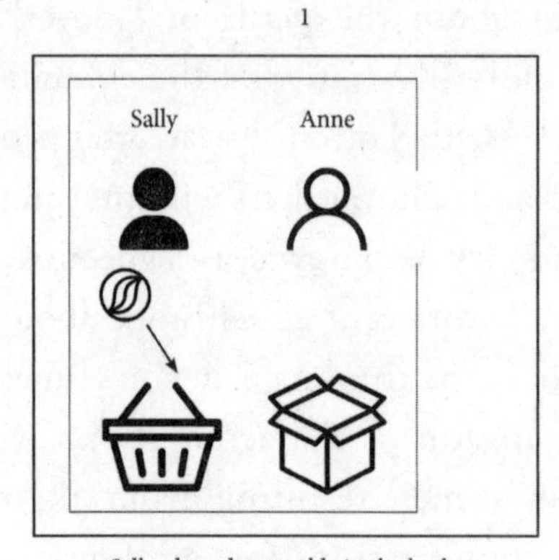

Sally places her marble in the basket

2

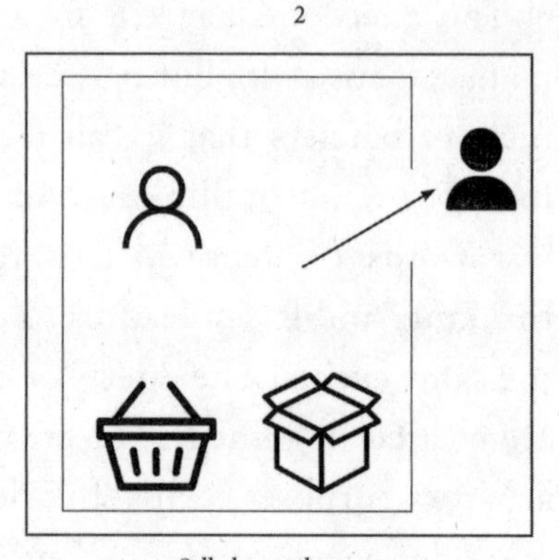

Sally leaves the room

3

Anne transfers Sally's marble to the box

4

Re-enter Sally. Where will she look for her marble?

Children under four years old tend to say that Sally will look for the marble in the box; since *they* know that the ball has been moved to this new location, they assume that Sally will too, without stopping to consider the fact that she had left the room when Anne played her trick. To solve the problem correctly, we must mentally place ourselves in Sally's shoes and recognise that without having seen Anne move the marble, she will expect to find it where she left it. This is no mean feat. Scientists have developed 'theory of mind' tests for many other animals, but few species manage to pass them.

Theory of mind – and the empathy that it engenders – is hugely important for creating a shared reality with others. It allows us to recognise when our beliefs are aligned with another person and when we might have to work harder to bring others around to our way of thinking. When it is functioning well, we can see that someone else is struggling and provide them with the support and validation they require; and when we require assistance, we can express our needs in a way that others will easily understand. Ideally, our theory of mind would always function with optimum accuracy, but like countless other complex abilities, it is not nearly as reliable as we might wish. In many situations, particularly those filled with ambiguities and distractions, we may simply forget to consider others' perspectives.[3]

If you think that you might be immune to these errors, consider the following theory of mind test – known as the 'director task' – that tends to trip up adults as well as children. Two participants are asked to sit at opposite sides of a bookcase populated with different objects. One is named a 'director'. Their job is to issue instructions to the 'addressee', who moves the objects manually. As you can see in the following diagram, some of the shelves are hidden behind screens, so the director can't see them, and the addressee needs to take this into account when executing their movements.

You can try it now. Place yourself in the shoes of the addressee, and imagine the person on the other side of the bookcase asked you to move the mouse. Which would you choose?

Someone attentive to the other person's point of view would know that it cannot be the toy mouse in the second row from bottom (which would be hidden from the person issuing the instructions). Instead, the director must mean the computer mouse, in the third row of the first column. That answer might sound obvious when we think it through logically. Yet many neurotypical adults struggle to take their partner's perspective into account consistently; on average, they reach for the obscured object about 25 per cent of the time.[4]

To delve further into the thought processes behind these decisions, researchers set up a video camera that tracked the movement of the participants' gaze as they made their decisions. From this, they could see that even when the participants successfully put themselves in the director's shoes, their eyes would often flick towards the wrong item before selecting the right object. 'Egocentrism isn't outgrown so much as it is overcome each time a person attempts to adopt another's perspective,' the authors concluded.[5] And in around a quarter of the trials, the adults failed to do that.

Psychologists describe this as a 'dual process' account of mindreading: there is an automatic impulse (egocentric thinking) that needs to be overridden with conscious deliberation (perspective taking). Since that second step is effortful and requires cognitive resources, we are more likely to fail when we are stressed or distracted, or when we don't care much about the consequences of our actions. People tend to think more egocentrically when they are put under pressure, for example. Increased incentives have the opposite effect. When they are financially rewarded for accurately perceiving another's viewpoint, people tend to think less egocentrically.[6]

In everyday life, we may not notice when we're slipping into that automatic, egocentric way of thinking. Without recognising our bias, we may ground our perceptions of others in our own experiences and fail to fully adjust for differences in knowledge or circumstance that might lead them to take very different views.

Consider a phenomenon known as the false consensus effect. In one iconic study, participants were asked to rate their agreement on a series of innocuous statements, such as:

- I like poetry
- I am embarrassed by dirty stories
- I have no fear of spiders
- I do not worry about catching diseases

People could strongly agree or disagree, or report feeling neutral, and those answers strongly influenced how common they believed the view to be across the whole population. People who liked poetry, for example, believed that they shared that taste with many others, while those who loathed it thought that most other people would find the verse similarly tedious.[7] Overestimating our similarity to the general population may soothe our fears of existential isolation, but it is an illusion, leading to nasty surprises when we find that our views are not as widely held as we had expected.

You might have noticed the false consensus effect when making new acquaintances. The talk seems to flow easily until one person suddenly drops an unexpected opinion that falls like a hand grenade into the otherwise pleasant conversation. The false consensus effect is also evident in discussions of moral values. Athletes who have taken performance enhancing drugs, for example, tend to overestimate how widespread the behaviour is among their competitors.[8] 'Everyone's doing it,' they assume – so why should they worry about the consequences of their behaviour? People who make racist, sexist or homophobic jokes do the same: they express their opinions as if they are commonplace, without considering that others may find their words deeply offensive.

Any overly confident view, breezily declared as if it were a truth universally acknowledged, can be highly jarring and shake the sense of shared reality for those who do not hold that opinion. It could be accidentally dissing someone's favourite band, deriding a political belief or talking immoderately about a mutual acquaintance. When we express a controversial opinion without even considering whether others might disagree strongly, there is likely to be at least a short moment of awkwardness. And in hindsight, we may wish that we had taken just a moment longer to think more carefully about the range of potential opinions before we spoke.

THE ILLUSION OF TRANSPARENCY

Besides creating false confidence in our ability to read *others'* minds, egocentric processing also leads us to overestimate the capacity of others to understand *our* internal life. This is known as the illusion of transparency. Thomas Gilovich at Cornell University has pioneered the scientific study of this phenomenon, which he and his colleagues compare to a scene from Edgar Allan Poe's 'The Tell-Tale Heart', in which the unnamed narrator has committed a murder and hides the dismembered body under his floorboards. When he is questioned by police, he believes that he can now hear the heart of his victim beating loudly. In reality, of course, it is the sound of his own pulse ringing in his ears. But the narrator does not recognise this fact, and, convinced that the officers must now be aware of his guilt, he confesses to the crime.

Like the failures in the classic theory of mind tests, the illusion of transparency arises from a failure to ignore our immediate perspective. Despite knowing – intellectually – that others cannot see inside our heads, we cannot fully overcome the intuition that our thoughts or feelings will 'leak out' in obvious ways. As a result, we believe that our emotional state is more obvious than it is.

The team's first experiment took the form of a lie detection contest. The participants took turns to make either a factual or fictitious statement about their lives while the others judged whether they were telling the truth or not. In line with multiple other studies of lie detection, the participants performed little better than chance at identifying the dishonest claims. When they were telling the untruth themselves, however, they assumed that their guilt would be written all over their faces – and so they significantly overestimated how accurately the others would be able to recognise their deception.

The second experiment was inspired by that most awkward situation at a dinner party, when we find the host's food to be unpalatable but struggle to eat it anyway. Try as we might to disguise our feelings, we are sure that our distaste will be transmitted to everyone around us. But is our discomfort obvious? To test this scenario in a laboratory, the researchers asked a group of food tasters to take sips from fifteen small cups. Five of the cups contained a foul-tasting brine solution used to pickle vine leaves, while ten contained cherry-flavoured Kool-Aid. No matter what they drank, the participants were asked to mask what they were feeling as a camera recorded their reactions. The films would then be shown to ten observers, who had to determine whether they had enjoyed each drink or found it disgusting. Thanks to the illusion of transparency, the tasters had little faith in their ability to disguise their displeasure. On average, they estimated that around half of the observers would have been able to recognise the feelings behind their masked expressions, whereas the true fraction was around a third.

The team's final experiment looked at moral discomfort. Participants were recruited for an experiment that ostensibly examined the influence of a work environment on problem solving. The researcher conducting the experiment pretended that this was the final part of his dissertation research, and so asked that everyone take the task as seriously as possible. Before leaving the lab, he selected one participant to lead the session. Their responsibilities involved writing the problems on the board and recording the other participants' responses. This person was not chosen randomly – she was a plant who had been told to break all the rules of the experiment. Rather than staying silent, she started giving people clues to the answers and, when judging their answers, frequently gave them credit for problems they hadn't fully solved. She even turned back the clock to give them extra time.

After the experiment, many of the participants reported feeling highly concerned about the assistant's behaviour. Despite not having said anything aloud, they assumed that their disapproval would have been evident to the people around them. When questioned, however, the participants failed to notice the others' discomfort; they assumed they were alone in their disapproval. From the outside, everyone appeared less alarmed than they felt inside.[9]

Knowledge of this illusion could initially come as some relief. It's useful, after all, to conceal the way we feel if our expressions of disgust are going to offend someone over their terrible cooking. Similarly, when feeling nervous about a job interview, we may be relieved to discover that our discomfort is not nearly as evident as we fear. Knowing about the illusion of transparency is yet one more reason to feel a little more confident about the way that we might be perceived. Along these lines, Gilovich has shown that educating people about the illusion of transparency can increase their confidence and improve their performance at public speaking.[10]

The illusion of transparency has many downsides, however. If we are buckling under our workload, for instance, we may hope that our boss can see when we are stressed and need help. Thanks to the illusion of transparency, our strain will be far less evident than we believe. If our boss fails to act, they aren't being unusually insensitive or uncaring; the great pressure we are experiencing and its effects on our mood may be occupying our minds but they simply aren't apparent to those on the outside.

It's not just negative feelings that will be hidden; the illusion of transparency could be masking our more pleasant emotions, too. Inside, we may feel a little leap of joy when we see our loved ones; we may believe our gratitude and affection are oozing out of every pore. On the outside, however, we may appear impassive and uncaring. We simply cannot assume that our feelings will have translated.

THE ILLUSION OF UNDERSTANDING

Our final illustration of egocentrism takes us to the limits of language and our ignorance of the ways our words may be misinterpreted.

Imagine that your friend Mark has gone to a new Italian restaurant, recommended by his colleague June, who declared that her dinner was 'marvellous'. When Mark eats there, he finds that the food and the service are distinctly mediocre. The next morning, he leaves a note on June's desk that reads 'You wanted to know about the restaurant, well marvellous, just marvellous.' Do you think that June would interpret this as sarcasm?

Having read about Mark's bad experience, we might guess that he was intending to convey irony. June, however, would be oblivious to that fact; based on her experience of a very pleasant meal, she would probably think he was being sincere. At the very least, we might declare that we are unsure how she would take it. To come to that conclusion, however, we have to discount our privileged knowledge of Mark's experience in the restaurant, and research by Boaz Keysar at the University of Chicago suggests that a sizeable number of people – around 60 per cent – fail to do this. They assume that June would be able to detect his sarcasm. Since they understand Mark's intentions, they assume that June would too, even though she was missing the necessary information to come to that conclusion.[11]

We can only conclude so much from people's assessments of fictional scenarios, but a later study shows that people often struggle to predict how their own writing will be interpreted. Each participant was asked to compose messages about ten different topics, with five serious sentences and five sarcastic sentences, which they would exchange with another participant. In each case, the participant had to predict how their partner would construe their statements. The participants expressed little doubt that any of their messages would be successfully decoded, but their partners misunderstood the intentions around 16 per cent of the time.

The same was true when they were asked to express emotions such as anger or sadness in their sentences: the writers' sentiments were not nearly as clear as they had assumed.

In my favourite of all these experiments, the researchers turned their attention to humour. Half the participants first watched *Saturday*

Night Live sketches, before sending out an email recounting one of the jokes. The rest simply read the jokes before sending out the email, without watching the sketch first. Thanks to their egocentric thinking, focused on their immediate experience, those who had already watched the comedian deliver the joke were considerably more likely to overestimate how funny the recipient would find their email. They forgot that as much of the humour came from the delivery – which was completely missing from the email – as the text itself.[12]

Even straightforward factual sentences can hide ambiguities that we often fail to recognise. Think about a simple sentence such as 'Angela shot the man with the gun'. We could read this as meaning either that Angela picked a gun out of a selection of weapons to shoot a man, or that she shot a man who was holding a gun. Either interpretation is possible, but most people tend to neglect this possibility and assume that others will come to the same conclusion as them.

In speech, we might hope that our vocal inflections would clear up the confusion – signals that are lost in text. People who recorded ambiguous sentences such as 'Angela shot the man with the gun' tended to assume that their intended meaning would be clear. When participants listened to those recordings, however, they were generally unable to guess which interpretation the reader was trying to convey.[13]

Such ambiguities can also be found in common phrases such as 'What have you been up to?', which could be heard as a genuine interest in someone's wellbeing, or the suggestion that someone has acted badly. We assume that our delivery will lead listeners to the right interpretation, but this doesn't happen nearly as often as we expect. The fact that *we* know what we are hoping to convey means that we assume the other person will too.[14]

The illusion of understanding, as this is known, can be a particular problem when we are conversing with someone of a different background from our own. I've often noticed this with American colleagues. Despite being familiar with the different cultural and linguistic norms, I often fail to remember how easily the subtext of British euphemisms can be misread by people who have not lived here. (If a Brit says 'that's a brave idea', it's probably not a sign of encouragement, for example.[15])

An Italian friend reports finding similar communication issues during his first years in London. He's often surprised by how reluctant British people are to accept casual favours, such as an offer to bring food when the other person is busy or ill. I had to explain that British people are often so worried about 'being a burden' that we may wait for the other person to explain how little they will be inconvenienced by the gesture, before we accept. Thanks to our egocentric thinking, it is easy to forget these differences – even after we have learned of their existence.

In its most extreme form, the illusion of understanding can lead us to ignore the language barrier when conversing with people from another country. Becky Ka Ying Lau and colleagues at the University of Chicago asked Mandarin Chinese speakers to record messages in their native tongue, which were then played to Americans who had no knowledge of the foreign language. The listeners had four multiple-choice options and had to pick the option they believed best represented what had just been said. Despite the obvious obstacles to interpretation, most of the people involved were overconfident in their judgements. The speakers believed that the listeners would be able to understand the gist of their words better than chance, and the listeners believed that they'd been able to pick out the correct meaning from the available options. Neither was true.[16]

As J. Edgar Hoover's command to 'Watch the borders' shows, the illusion of understanding can sometimes have dramatic consequences. In their paper, the researchers at the University of Chicago describe how the communications office had once received an urgent message from the university hospital's PR office warning that there was a 'shooting on campus'. After some panic, the staff realised that the media officer had been referring to a movie that was being recorded on the campus grounds: it was a *film* shoot. Owing to the illusion of understanding, he had assumed that his message was loud and clear – yet it had nearly resulted in a panic.

The illusion of understanding can even be life threatening. Doctors must pass on case details about their patients to their colleagues on the next shift – but detailed interviews with those concerned show that the handover is often a hotbed of confusion. In one study of paediatric interns, Keysar and colleagues asked each physician about the information they wished to convey and then compared this to the understanding of the next doctor on the ward. They found that the most important detail was misunderstood by the second physician around 60 per cent of the time – despite the first doctor having believed that they'd communicated the facts (and their importance to the patient) effectively.[17]

We may not hold others' lives in our hands, but in our everyday relationships, the illusion of understanding can lead to considerable frustration as we fail to make our meaning clear. We are then baffled to find others behaving in a way that appears to be at total odds with the words we said. Just think how many arguments would be solved if we simply made an extra effort to check each other's interpretation of the points under discussion.

THE CLOSENESS-COMMUNICATION BIAS

By September 1999, a spacecraft called the Mars Climate Orbiter had spent nine months travelling more than 669 million kilometres across space towards the red planet. Every step of its journey had been meticulously choreographed through detailed coordination with ground control, and on the twenty-third of that month, it was all set to make one last pirouette into Mars's orbit. By 9.04 that morning, however, NASA's scientists had lost contact. They never heard from the spacecraft again. In the days after the disaster, the world's media were curious to know how such a promising mission could have ended so miserably – and the answer was basic miscommunication. The software on ground control had been sending instructions using imperial measurements, while the software on the spacecraft was operating in metric. The resulting miscalculation led the Orbiter to veer from its precisely calculated path into the planet's atmosphere, where the friction of the air caused it to crash and burn.[18]

The repercussions of that mistake may have taken place millions of kilometres from Earth, but it was an all-too-human error. The team responsible for one piece of software had not checked in with the team producing the other about this crucial detail. They had simply assumed that they were all working from the same frame of reference.

Knowing about our egocentric thinking and the illusions of transparency and understanding, I can't help but think that the spacecraft's fate is a perfect analogy for many of the missteps we make navigating the social interactions with the people in our orbit. We may do everything we can to open the way for friendship, only to use the wrong frame of reference to express our thoughts and feelings and to interpret those of others. These errors do not always end in catastrophe, but they prevent us from connecting in the ways that we'd hope.

The opportunities for miscommunication do not decrease on acquaintance. In the director task, for instance, participants took longer to correct their egocentric thinking when they were taking orders from a friend than a stranger, and were more likely to make the error of reaching for the wrong object. And two friends or two spouses are just as likely as two strangers to misunderstand the meanings of ambiguous words and turns of phrase. The big difference was their complacency: when questioned about the likelihood of making errors, the friends and spouses were even more likely to overestimate their capacity to communicate successfully than the strangers were.[19]

This phenomenon is known as 'closeness-communication bias' and it is a major challenge in maintaining our connection. Having established a shared reality over certain kinds of experiences, we may quickly come to assume that we will always agree on every point. We therefore forget that in certain situations, we still need to make a conscious effort to build mutual understanding. We may have the same music and literary taste, or share the same moral values – things that increase our feelings of closeness – but our friends or partners may still need us to explain why we hold a particular opinion or behave a certain way. Two minds are unlikely to work in perfect synchrony all the time – and if misunderstandings, however small, are left unresolved, they will remain an awkward wedge between us.

The latest research suggests that overconfidence in our perceptions of the people around us is extraordinarily prevalent in many domains. People consistently overestimate their capacity to guess their partner's preferences for different hobbies, their personal dreams and their opinions on topics such as policing.[20] No matter how well we know someone, there will always be new depths to discover, and it is far better to acknowledge this fact than to proceed with false assumptions.

OVERCOMING EGOCENTRISM

If you are a fan of self-help books such as Dale Carnegie's *How to Win Friends and Influence People*, you may assume that you can overcome egocentric thinking through the power of imagination. Simply think more carefully about the other person's situation and try to see the world from their point of view. Like most ideas in folk psychology, the idea of 'placing yourself in another's shoes' has the ring of truth, but is it as powerful as we have been led to believe?

Deliberate perspective-taking could certainly correct some of the egocentric errors that we have encountered. From the false consensus effect to the verbal illusions of understanding, it can be useful to think a little more carefully about what the other person does or does not know. Such efforts are often rewarded with greater connection: simply hearing that someone has made the effort to take our perspective increases feelings of closeness, measured by people's responses to the Inclusion-of-Other-in-Self scale (see page 37).[21] We seem to appreciate any sign that someone is trying to reduce the distance between us and understand our point of view.

Some people are naturally more motivated to practise perspective-taking than others. On a scale of 1 (does not describe me well) to 5 (describes me well), how would you rate the following statements?

- I believe there are two sides to every question and I try to look at both of them
- I sometimes try to understand my friends better by imagining how things look from their perspective
- Before criticising somebody, I try to imagine how I would feel if I were in their place

These are items from a psychological scale measuring 'trait perspective-taking', and higher scorers tend to enjoy smoother social interactions with fewer conflicts and misunderstandings.[22] Trait perspective-taking also predicts greater relationship satisfaction in married couples.[23]

Perspective-taking cannot work miracles, however, as Tal Eyal at Ben Gurion University and colleagues powerfully demonstrated in a study from 2018. The participants were asked to predict whether another person would like a particular movie, joke, work of art or leisure time activity. In some cases, the other person was their boyfriend or girlfriend; in others, it was a stranger whom they had got to know through a written introduction and short conversation. In each case, however, they were told to put themselves in the shoes of their partner and carefully consider things like their personality, background and tastes. 'Imagine what they would like and dislike about each [activity etc.], and consider how that would influence their ratings of each [activity etc.].'

Eyal's surveys found that most people had great faith in the technique. When asked about the best strategy to arrive at mutual understanding, around 70 per cent believed that they could predict the insides of someone's head using mere imagination. Perhaps they were thinking of famous detectives, like Sherlock Holmes, who have such an astute sense of personality that they can predict whole patterns of behaviour from a handful of clues.

Sadly, such feats are only possible in fiction, and for most participants the accuracy of perspective-taking was distinctly disappointing. Indeed, in some cases, the attempts at perspective-taking led to considerably worse judgement, compared with those who were simply relying on their intuitions.[24] Their imaginative efforts may have been built from flawed sources of information – if we do not know the person, we may be relying on crude stereotypes to guess their thoughts and feelings; if they are already an acquaintance, we may be relying on a memory of an experience that is no longer relevant.

To fully overcome our egocentrism, then, we need to become versatile and flexible thinkers. If we have no better information to go on, it makes sense to use our past knowledge of someone's behaviour to guess their thoughts and feelings through perspective-taking. But we need to be humble about the accuracy of these assumptions, rather than talking or acting with complete certainty. And when the appropriate opportunity comes around, we should be ready to check our understanding with the other person and update our beliefs accordingly. No matter how empathetic we think we are, there is no better way of finding out what someone thinks and feels than simply asking them.

Eyal's team describes this as 'perspective-getting', as opposed to 'perspective-taking'. Don't be deceived by the simplicity of this distinction: it may seem obvious but it is often overlooked. By underestimating the benefits of asking for verification, we are missing many opportunities to clear up confusion that might drive a wedge in our relationships.

If we use people's answers as feedback, we may naturally train our theory of mind to be more accurate and sophisticated, so that we become more sensitive to potential misunderstandings before they have a chance to drive a wedge between ourselves and others. But that learning curve can only start from a place of humility.[25] To close the distance between ourselves and others, we must be constantly curious, and we must be ready to be surprised by what we find out.

* * *

In addition to correcting our own mistakes, this new knowledge of egocentric thinking can help us to be a little more sympathetic to others' errors. As we learned in the first chapters of this book, there are magical moments when others sense exactly what we are thinking and feeling, when someone knows exactly what we need – whether that's offering a hug when we're feeling down, picking the perfect birthday present or making plans for a perfect date. We cherish those occasions when they occur purely by chance and take them as a sign of our deep connection.

When that understanding does not happen spontaneously, we can feel frustrated and hurt, particularly if we've relied on that person for support in the past. We may even see these situations as a test of our relationship. We assume that the other person should know what's on our mind – and if they aren't able to appraise the situation, they can't care about us as we thought they did.

I've certainly been guilty of this myself. When people have misunderstood me or overlooked my emotional state, I've assumed they simply aren't making enough effort to see my perspective. I could, of course, have simply told them how I was feeling, but I had the strange notion that if I expressed it verbally, that would somehow diminish the value of their support. I wanted it to come of its own accord, without my intervention.

I now realise that this may be placing unrealistic expectations on the other person's mindreading abilities. Thanks to my own illusion of transparency, I may have not provided them with enough information to enter my inner world, while believing that my feelings were plain to see. I was judging *them* for *their* lack of empathy when the fault may have been entirely my own.

In its pithiest form, our fourth law of connection can therefore be stated as follows: **Check your assumptions; engage in 'perspective-getting' rather than 'perspective-taking' to avoid egocentric thinking and misunderstandings.**

Whether we have just met someone, or have been friends with them for years, there will always be new opportunities to build on our acquaintance and get to know each other better. Each person is a multidimensional being with an infinite capacity to surprise, and we may all build much stronger relationships by respecting this simple fact and allowing ourselves more time and effort to discover our similarities and understand our differences.

What you need to know

- Theory of mind describes our ability to predict others' mental states – including their beliefs, intentions and emotions. Most neurotypical adults have basic theory of mind, but it is unreliable
- When we are under pressure or distracted, many of us are swayed by egocentric thinking, in which we assume that others' mental states and perspectives are similar to our own and fail to take into account their particular situations and experiences
- Egocentric thinking can lead to ingratitude, as we fail to see the hidden efforts others have made on our behalf. It is also responsible for the false consensus effect, in which we assume that our beliefs and values are more common than they really are
- The illusion of transparency is the false assumption that our emotions and intentions – so salient in our own minds – will be easily visible to others. Its counterpart is the illusion of understanding, which leads us to overestimate others' capacity to understand the nuances in our writing and speech, while overlooking the ambiguities of the language we are using

Action points

- When you wish to convey an important message, try to cultivate a habit of thinking about the ways your messages might be interpreted. Double check whether you have included all the necessary information for the recipient to understand your thoughts. Consider their experience in the matter at hand and the context in which they will be reading or hearing it
- When you want others to understand the way you feel, make a little more effort to articulate your emotions and your reasons for

experiencing them, rather than assuming that they will be evident from your facial expressions or body language

- Feel free to practise perspective-taking, but be aware that your predictions of others' beliefs, values and experiences may be mistaken, and look for ways to confirm or update your assumptions. Simply asking for more details is the best way to build an accurate understanding of another person's inner life – yet few of us do it naturally

CHAPTER 5

THE ART OF CONVERSATION

If you've ever spoken to someone and later felt that you would have better spent your time talking to a brick wall, you'll surely identify with the observations of Rebecca West. 'There is no such thing as conversation,' the novelist and literary critic wrote in her collection of stories, *The Harsh Voice*. 'It is an illusion. There are intersecting monologues, that is all.'[1] West goes on to describe each person's speech as circles, emanating from the speakers' mouths, that simply pass each other by without touching; neither mind is changed by the encounter. Her sentiments recall Sylvia Plath's fears that she would never be able to relate to another person. If someone feels that their conversations have left no impression on those around them, then that is the definition of existential isolation. You may not have known the term, but you've probably experienced this yourself on a bad date, at an awful dinner party or during an interminable family gathering.

The illusions of transparency and understanding that we explored in the last chapter can offer one explanation for the breakdown of our conversations into 'intersecting monologues', but these are only the beginning of our communication problems. Psychological research has identified many other habits and biases that impose barriers between ourselves and others – and if we wish to have greater connection with the people around us, we must learn how to overcome them. The good news is that these corrections are all very easy to put into practice. From our choice of opening gambits to our expressions of interest in the other person's life and the manner in which we describe our own experiences, tiny tweaks to our conversational style can bring enormous benefits.

HAZLITT'S LAW

Let's begin with the sins of inattention. 'The art of conversation is the art of hearing as well as of being heard,' declared the early nineteenth-century essayist William Hazlitt in his 'On the Conversation of Authors', published in 1820. 'Some of the best talkers are, on this account, the worst company.'

Hazlitt noted that many of his literary acquaintances – who included Samuel Taylor Coleridge, Stendhal and William Wordsworth – were so keen to show off their wit and intelligence that they lacked the basic civility of listening to others. He instead recommended that we imitate the painter James Northcote, who, he claimed, was the best listener and – as a result – the best converser that he knew. 'He lends his ear to an observation as if you had brought him a piece of news, and enters into it with as much avidity and earnestness as if it interested himself personally,' Hazlitt wrote. 'I never ate or drank with Mr Northcote; but I have lived on his conversation with undiminished relish ever since I can remember,' he continued. 'And when I leave it, I come out into the street with feelings lighter and more ethereal than I have at any other time.'[2] Who wouldn't want to leave their acquaintances feeling this way?

The simplest way of achieving this effect is to ask more questions. You may have heard this advice before – it's an elementary social skill – but it bears repeating, since recent research suggests that surprisingly few people have cultivated this habit effectively. While studying for a PhD in organisational behaviour at Harvard University, Karen Huang invited more than 130 participants into her laboratory and asked them to converse in pairs for a quarter of an hour through an online instant messenger. She found that, even in these fifteen minutes, people's rates of question-asking varied widely, from around four or fewer at the low end to nine or more at the high end. And asking more questions can make a big difference to someone's likeability. In a separate experiment, Huang's team analysed recordings of people's conversations during a speed-dating event. Some people consistently asked more questions than others, and this significantly predicted their chance of securing a second date.

Given shared reality theory, it's easy to understand why questions are so charming: they demonstrate your wish to build mutual understanding and give you the chance to validate each other's experiences. Consistent with the other research on our social biases, however, Huang found that most of her participants overlooked these benefits. In one of her studies, participants were presented with various potential strategies to handle a forthcoming conversation, one of which involved asking many questions, while another involved asking few. Hardly any of the participants appeared to recognise the advantages of the first option, despite it representing one of the most powerful ways to make a connection.

Even if we do pose lots of questions, we may not be asking the right kind. In her analyses, Huang considered six different categories of questions. You can see the examples below:

Introductory	Hello! *Hey, how's it going?*
Follow-up	I'm planning a trip to Canada. Oh, cool. *Have you ever been there before?*
Full switch	I am working at a dry cleaner's. *What do you like doing for fun?*
Partial switch	I'm not super outdoorsy, but not opposed to a hike or something once in a while. *Have you been to the beach much in Boston?*
Mirror	What did you have for breakfast? I had eggs and fruit. *How about you?*
Rhetorical	Yesterday I followed a marching band around. *Where were they going?* It's a mystery.

Huang found that the follow-up questions, which ask for more information about a previous point, are much more appealing than the 'switch' questions that change topic, or the 'mirror' questions that simply copy what someone has already asked you. The most superficial are the introductory questions. These icebreakers may be essential social niceties, but they hardly demonstrate a genuine interest in another person, and they are unlikely to elicit the kinds of details that could help you both to build a shared reality. If they are the be-all and end-all of our attempts to draw out another person – and they are, for many people – then our conversational style may need an overhaul.[3]

You might also avoid boomerasking – that's the habit of posing a question as an excuse to talk about yourself. We could ask about someone's profession, for example – not because we care how their job is going, but because we want to brag about our own promotion. Huang did not consider this as a separate category in her analyses, but emerging research suggests that this habit is particularly unlikeable, contrary to boomeraskers' assumptions that they are being inclusive and curious.[4]

The act of asking elaboration questions can become self-perpetuating. Once someone had made the effort to draw out their partner with one enquiry, it became much easier to ask another. You will find new avenues of conversation opening in many different directions – each of which can blossom into a new opportunity for engagement and social connection.

THE ART OF ATTENTION

There is much more to social curiosity than asking the right questions. People are acutely aware of whether they are being listened to attentively, and their perception of receiving active attention from another predicts their feelings of trust towards that person, and contributes to the wellbeing boost that typically comes from strong social connections. The formula is simple: the more attentive we are to someone, the happier they feel.[5] In the workplace, the enhanced feelings of trust and safety that come from being heard and understood can even render us more creative.[6] Unfortunately, many of us rely on the wrong cues to signal our interest in others.

People can display their attention with non-verbal body language, such as leaning forward in their chair, nodding their head, or making empathetic facial expressions; they can employ 'paralinguistic' cues such as murmuring sounds of assent or approval; or they may verbally acknowledge what the other person has said. While non-verbal and paralinguistic cues are often genuine signs of attention, they can also be feigned while our mind is wandering – and if we rely on these alone, our conversation partners may often assume the worst. As we saw with the studies of the liking gap, we tend to wallow in these kinds of doubts after a conversation has ended, and that can erode our feelings of connection.

For this reason, social psychologists argue that it is much safer to demonstrate your attention *explicitly* in the words that you say.[7] Paraphrasing what the other person has said, for example, offers direct proof that you have absorbed and processed their remark: there is no way of faking that response. This is another reason why follow-up questions are so powerful: the details that you include provide the necessary confirmation that you were more intent on hearing what they had to say than on being heard yourself.

Be careful to focus on the core point that the person has been trying to convey, rather than some circumstantial detail. If someone describes a bad date to you, for example, it's no good enquiring about the bar or giving your opinions on the film that they watched, rather than discussing their feelings of disappointment or frustration with their prospective partner.[8] Joint attention is one of the central characteristics of having a shared reality, after all; if you both are processing the world in the same way, you should be focused on the same elements. You can validate what they are thinking and feeling or perhaps, after acknowledging what they have said, offer an alternative interpretation that may open their mind to a new way of seeing the situation. You don't always have to agree with them – but you must show that you are at least trying to see things their way before offering your alternative take on the situation.

As you converse, avoid being distracted by your surroundings. On the day of starting this chapter, I was invited to meet up with a former colleague for coffee. During our chat, however, I couldn't help but notice that his eyes were constantly wandering around the room and made contact with almost any woman who walked past. Perhaps not coincidentally, he told me he was on the verge of splitting up with his long-term girlfriend. It was highly distracting, for me, to see his mind wandering so wildly, and although the conversation covered plenty of interesting topics and we found many points of agreement, I left the café feeling that he might as well have been talking with a chat-bot. You might observe similar behaviour at a party, when your conversation partner is constantly looking over your shoulder to catch the attention of other people that they know. Some disruptions of this kind will be inevitable, but each time that you show your mind is wandering, you weaken the connection that could have arisen from more attentive listening.

The practice of 'phubbing' – phone snubbing – is similarly disruptive. This is the act of constantly interrupting a conversation to check your smartphone for new notifications – and it damages rapport. In one observational study, researchers watched 100 pairs of participants conversing in local coffee shops. Some participants naturally took out their phones and held them in their hands or placed them on the table, while others left them out of sight. At the end of the conversation, the researchers asked each person to fill out a questionnaire exploring the experience, and they found that the mere presence of the phones on the table reduced the pair's feelings of empathy for each other, resulting in a less fulfilling conversation.[9] The researchers call this the 'iPhone effect' – and the result was later replicated in a randomised experiment, in which participants were instructed to keep their phones either hidden or on the table as they talked. As predicted, the distraction of having the phone in sight decreased the perceived quality of the interaction, while those who had taken their mobiles out of their eyeline tended to have more rewarding conversations.[10]

When mastering the art of attentive listening, it is worth remembering that some rules are made to be broken. As a child, you were almost certainly told not to interrupt your elders and betters, and this is common advice in many etiquette guides – breaking someone's train of thought, midway through a sentence, is considered egregious manners. This is undoubtedly true for the many occasions in which people interrupt others' discourse to direct attention back to themselves, but there are important exceptions. When you already feel truly connected to the other person, you may find that your thoughts are running in the same direction and that you are in total agreement over the issue at hand. In these situations, a gentle interjection to complete their argument can underline your joint understanding and the sense that your minds have aligned.

We've already seen one example of this; you may remember that in Maya Rossignac-Milon's experiments, participants who finished each other's sentences and spoke at the same time as each other tended to score more highly on the Generalised Shared Reality scale.[11] And Daniel McFarland at Stanford University found exactly the same results in his recordings of speed daters' conversations: people reported clicking *more* with the people who interrupted them. Looking more closely at the data, he found that the interruptions were 'collaborative completions' that helped to fill in the gaps of someone's argument when they were struggling to articulate what they meant, such as:

Female: So are you almost–

Male: On my way out, yeah–

Or they validated the other by emphasising a shared opinion or perspective, such as:

Female: [*laughter*] Yeah. Yeah, I went there last year for Radiohead because they're like my all-time–

Male: Oh yeah, Radiohead are great.

Or:

Female: Yeah and listen, not devoting your full attention. Kind of like–

Male: [*laughter*] Exactly. Yeah I had those thoughts.[12]

Our natural dispositions may influence how well we employ these kinds of interjections. If you too rigidly stick to the rules of good manners, you might studiously avoid such displays of enthusiasm, even though they would help to build a bond. This can be a particular problem for socially anxious individuals, who may be so preoccupied with the other's judgement that they fail to jump in at the right moments, and remain too passive participants in a conversation.[13]

Socially confident extroverts may face the opposite problem. In a series of studies at Stanford University, people showing high levels of extroversion were consistently seen as less attentive conversation partners. This seems to arise from the sense that extroverts are less authentic in their presentation; their enthusiastic and energetic behaviour is interpreted as a form of acting, which hides the fact that they were not really paying the attention and respect that we desire. While you are talking about a recent trip, for example, an extrovert might excitedly exclaim, 'Wow, that's so interesting!', which is certainly better than ignoring the point and shifting the topic to oneself, but could easily have been faked – unless they then follow up with more questions or validate your experience with more in-depth observations about what you have just said.[14]

Introvert or extrovert, we should always strive to express our attention and interest in the most specific way possible, by focusing on the particular details of what the person says and making a conscious effort to deepen their understanding of the topic under discussion.

THE FAST FRIENDS PROCEDURE (AKA THIRTY-SIX QUESTIONS TO FALL IN LOVE)

Given Hazlitt's Law, we might conclude that we should always allow our acquaintance to take centre stage. They occupy the conversational spotlight while we sit in the wings; our role is to prompt new opportunities for them to elaborate on their thoughts and feelings, and then validate what they have said, without disclosing much about ourselves.

This advice can be found in many influential etiquette guides, but psychological research shows that it is misguided: we should feel free to take our fair share of the airtime.[15] The creation of a shared reality between two people relies on us understanding *each other* – and that's not possible if we only take on the role of the listener without saying anything about ourselves.

For this reason, we should try to create conversations that allow both parties to open up about our deeper thoughts and feelings so that we can identify the points of common ground that may lie behind any superficial differences. Arthur Aron – who pioneered the theory of self-expansion that we examined in Chapter 2 – has powerfully demonstrated the advantages of self-disclosure, using an experimental paradigm that is sometimes known as the 'fast friends procedure'. The participants were first sorted into pairs. 'Your task, which we think will be quite enjoyable, is simply to get close to your partner. We believe that the best way for you to get close to your partner is for you to share with them and for them to share with you.'

They were then given a series of questions to discuss over the next forty-five minutes.

To compare the effects of high and low self-disclosure, Aron prepared two different sets of discussion points. Half the pairs saw questions that stimulated small talk, such as:

- How did you celebrate last Halloween?
- Describe the last pet you owned
- Where did you go to high school?
- Do you think left-handed people are more creative than right-handed people?
- What was the last concert you saw? How many of that band's albums do you own? Had you seen them before? Where?

This was the low self-disclosure condition. They were perfectly reasonable questions – the kind you might happily ask on a first date – and the answers might prompt some pleasant discussions. But they weren't necessarily going to provide many profound insights into someone's inner life.

The rest of the participants were asked to discuss more probing questions, such as:

- What would constitute a perfect day for you?
- If you were able to live to the age of ninety and retain either the mind or body of a thirty-year-old for the last sixty years of your life, which would you want?
- Do you have a secret hunch about how you will die?
- If a crystal ball could tell you the truth about yourself, your life, the future or anything else, what would you want to know?
- What, if anything, is too serious to be joked about?
- Your house, containing everything you own, catches fire. After saving your loved ones and pets, you have time to safely make a final dash to save any one item. What would it be? Why?

This was the high self-disclosure condition. The aim was to get the pairs to open up to each other about their specific thoughts and feelings, with answers that more directly reflected the idiosyncrasies of their minds. In each case, the participants were asked to engage equally. 'One of you should read aloud the first slip and then *both* do what it asks,' they were told.

After the forty-five minutes were up, the participants were asked to describe how close they felt to the partner, with questions like the Inclusion-of-Other-in-Self scale. As you may remember, this presents a series of increasingly overlapping circles labelled from 1 to 7 and asks participants to state which picture best describes their feelings towards the other person (see p. 37).

In this first experiment, the people in the high self-disclosure condition rated their relationship as 4, while those in the small talk condition rated themselves as 3. This would be a relatively large effect size for any single psychological intervention, but it's especially noteworthy when you consider that most people's lasting friendships do not score much higher. Indeed, when Aron compared the Inclusion-of-Other-in-Self ratings to studies of students' broader social networks, he found that these new relationships were rated as being more intimate than many people's closest relationships.[16] (In the media, Aron's experimental procedure is often labelled as 'thirty-six questions to fall in love', though the study did not explicitly examine romantic desire.)

The benefits of self-disclosure, through the fast friends procedure, have now been replicated in large studies and they have shown that it is just as effective during remote communication as face-to-face interactions.[17] Researchers at the University of Hagen – an institution for distance education in Germany – created an online version of the task for 855 remote learners enrolled on the undergraduate psychology degree. As hoped, the fast friends procedure increased the feelings of social connection between the (virtual) classmates and ensured that more of the students continued the course until the final exam, rather than dropping out.[18]

Self-disclosure can even increase connection among people from different social groups, reducing the prejudices and suspicions that are typically tied to an 'us-and-them' mentality. The team at Hagen, for instance, found that the procedure increased the participants' sense of closeness regardless of differences in demographic factors, such as age or immigration status, that you may expect to pose as barriers to friendship. Along similar lines, scientists at Stony Brook University in the US have shown that the procedure helps foster social connection between people of different sexual orientations; after going through the thirty-six self-disclosure questions with a gay or lesbian participant, straight people revealed less prejudiced attitudes on a survey, and greater feelings of closeness towards that person.[19] The procedure has also been enrolled to encourage interracial friendships among middle-school students, with some success.[20]

If you feel uncomfortable at the thought of self-disclosure, you are not alone: many people fear exposing themselves. When asked to predict how they will feel during the exchange, most people expect that the fast friendship procedure will be painfully awkward, and they struggle to see how it will lead to feelings of closeness; they expect small talk to be far less painful. When they engage in the task, however, the conversation flows far more smoothly than they expected, and afterwards they report feeling a greater sense of connection with their conversation partners than they had thought possible. As a result, they are considerably happier than they had foreseen. [21]

One major psychological barrier is the anticipated interest: people expect their partners to be indifferent to them and to be bored by their self-disclosure. But the research on the fast friends procedure suggests this is not the case: people are far more interested in our innermost thoughts and feelings than we imagine. Self-disclosure requires a leap of faith, but when we make it, we tend to land safely.

People who have undertaken the fast friends procedure and engaged in heightened self-disclosure begin to show some of the physiological markers of social connection. You may remember that when we form a shared reality with someone, our brains and bodies begin to synchronise as we both read and respond to the world in the same way. Our hormonal responses to stress become attuned, for example – so that levels of cortisol rise and fall in tandem as we experience the same emotional responses to events. This phenomenon has been documented in friends, lovers and family members – and a recent study from Columbia University shows that it also occurs among people who have just taken the fast friends procedure together. People who engaged in more banal small talk do not synchronise in the same way.[22]

The warm feelings of affection and trust that arise from self-disclosure seem to be aroused by the release of natural opioids in the brain, which encourages further bonding.[23] As the name suggests, this class of chemicals includes drugs that share the same structure as opium, such as morphine, and in 2019 Canadian researchers showed that the enhanced social connection emerging from the fast friends procedure arises through the same mechanism. To prove this, the scientists turned to a drug called naltrexone that blocks the brain's opioid signalling. This means that someone who is given morphine after taking naltrexone won't feel the expected pain relief or the sense of bliss that typically accompanies the drug. If the release of opioids can explain some of the buzz we get from social connection, then participants who have taken naltrexone should not reap such large benefits from the fast friends procedure.

To find out whether this really were the case, the researchers recruited around 160 participants who were divided into pairs. Half were given naltrexone, and the others a placebo, before they each discussed the thirty-six self-disclosure questions designed by Aron. After their chat, each participant took a series of questionnaires describing how the conversation had evolved. As expected, the participants who had taken the naltrexone – and therefore lacked the benefits of their endogenous opioids – were less open in the conversations; the relationship just didn't seem to be evolving in the same way, and it blunted the mood boost people normally experience following the exchange.

Besides helping pairs of conversation partners, self-disclosure may increase the intimacy of bigger groups. Along these lines, Robert Slatcher, then at the Wayne State University, tested the fast friendship procedure for pairs of romantic couples on a kind of double date. As expected, each couple feels closer to the other pair after they have discussed the thirty-six questions. Crucially, however, the individuals within each couple also feel greater love for one another. One reason may be that the new acquaintances helped to validate their relationship and increased confidence that their love was on the right track. Making friends with other people would have also provided each couple with a sense of self-expansion. They had the sense that they were growing together, which we know is important for relationship outcomes. All parties, in other words, were relishing the thrill of finding new people, with unique viewpoints, who are willing to open up and work towards creating a shared reality together.[24]

I've included a link to the full list of the thirty-six questions in Aron's fast friends procedure in the Further Reading section at the end of the book – if you ever happen to have the opportunity to use them, I'd heartily recommend it. Needless to say, this must be conducted with tact and discretion. While you might slip one or two into a conversation at suitable moments, you would look a little odd if you tried to memorise the whole list and rolled them out whenever you met a new acquaintance – unless, of course, you explain what you are doing. This isn't such a bad idea, particularly if you are talking to people with an interest in the human mind. I've found that talking about the fast friends procedure can itself be a good conversation starter. More importantly, however, you should draw on the spirit of this research by being a little more transparent about your deeper thoughts and feelings – and by giving others the opportunity to do the same. Whether you are describing a secret dream, expressing an unexpected emotional reaction to a news story or talking about a particularly precious memory – be generous with the information that you provide.

Eschewing small talk in favour of deeper conversations should boost your long-term life satisfaction. Researchers recently equipped 486 participants with a small 'electronically activated recorder' that allowed the scientists to eavesdrop on the participants' interactions throughout the day. Comparing the quantity and quality of the conversations with a standard measure of life satisfaction, the scientists found that the amount of time someone spent in small talk about daily banalities made almost no difference to their contentment, whereas deeper conversations involving the exchange of meaningful information about their circumstances and interests had a significant impact.[25] When you bare your soul, others will often respond in kind – and you will all feel better for it.

THREATENING OPPORTUNITIES

Given the power of self-disclosure and open dialogue, it's worth pausing here to examine the diversity of our friendships. The research on the fast friends procedure shows us that self-disclosure can instantly increase the rapport between people of different backgrounds, but we might still worry about discussing our differences head-on.

The psychologist Kiara Sanchez describes these interactions as 'threatening opportunities'. Working with colleagues at Stanford University, she has interviewed Black participants about their friendships with White people, for instance. Many hoped that greater closeness would emerge from open conversations about race, but they also feared being misunderstood. 'Race is a sensitive subject because no one wants to be offensive or be offended. Once that line is crossed it seems to define who you are as a person and [be a] cause for extreme scrutiny,' one participant told the researchers. Despite these concerns, most participants wished to disclose their experiences to their White friends: they expected the benefits to outweigh the risks. Interviews with White participants, meanwhile, revealed a genuine desire to listen and learn from their Black friends about their experiences of race; they really wanted to have greater understanding, but just didn't know how to broach the topics.[26]

We can see similar reports among LGBTQ+ people. You may think that their primary concern is explicit or implicit prejudice, but research by sociologists Rin Reczek and Emma Bosley-Smith shows many people's sense of alienation arises from a lack of overt interest. Their loved ones can be so afraid of causing offence that they avoid any discussion of the person's identity and the challenges they face.[27] The same will apply to matters of religion, disability, or neurodiversity: we can find our bonds strengthening if we discuss our differences openly.

Tact and sensitivity will be necessary with all kinds of self-disclosure, and we must ensure that our curiosity does not itself become a burden with overly intrusive questions. In some cases, it is best to offer the space for these discussions to occur when the opportunity comes up and let the other person know you are willing to open those conversations whenever they wish. And you should pay them the respect they deserve with attentive, mindful listening.

In her scholarly paper on threatening opportunities,[28] Sanchez quotes a line from the philosopher and writer Audre Lorde that perfectly captures the possibilities created by more open and understanding conversations between people of different backgrounds. 'When we define ourselves, when I define myself, the place in which I am like you and the place in which I am not like you, I'm not excluding you from the joining,' she wrote. 'I'm broadening the joining.'

THE NOVELTY PENALTY

Before we end our discussion on the art of conversation, we must look at one final psychological phenomenon known as the 'novelty penalty', which can be a barrier when discussing unfamiliar experiences.

The term comes from an experiment by Gus Cooney – one of the researchers who discovered the liking gap. His team first placed the participants into groups of three. While alone, each member watched one of two short videos: a TED talk about the intelligence of crows or an interview with the owner of a specialist soda shop. The trio then met as a group, and one member – the speaker – was asked to describe the video he or she had seen, while the other two members listened for two minutes. In some groups, the listeners and speakers had all watched the same video, while in others, the speaker spoke about the clip that the listeners hadn't seen.

You would expect that learning something new would be far more enjoyable and interesting than hearing something that is already known to you. But the listeners in Cooney's study had exactly the opposite reaction: they tended to prefer hearing about the same video they had just seen, while remaining underwhelmed by the talk that contained fresh information. This is the novelty penalty: a general preference to hear about familiar experiences over unfamiliar experiences.[29]

You may not have known its name, but you will have almost certainly noticed the novelty penalty when you have returned from an exotic holiday. Your mind is still full of all the sights, sounds, smells and tastes of the places and the amazing people that you met along the way. As you try to describe the experience, however, you may find people's eyes glazing over. It's not that your audience doesn't care. They simply don't have enough knowledge to immerse themselves in your descriptions and understand why the trip was so special to you. The informational gaps could create a feeling of distance that undermines the sense of a shared reality – compared to conversations about more familiar topics.

The novelty penalty might explain why writers since the time of Hazlitt have decried people who talk for too long about themselves; it can prevent self-disclosure from fostering connection, even if the topic of the conversation seems fresh and interesting. If you make sure that your words are always relatable to the other person, you can even get away with taking up more than 50 per cent of the chat.[30] When the other person has no way of linking what you say to their own experiences and knowledge, however, then even relatively few words can outstay their welcome.

One strategy to avoid the novelty penalty would be to focus your interactions on topics that are equally familiar to both parties, and to shun anything that falls outside their conversational comfort zone. This is certainly worth considering if your primary motive for raising an esoteric subject had been to impress the other person. You may think that it's cool to talk about music that no one else listens to, or films no one else has seen – but because of the novelty penalty, this can have the very opposite effect to the one you intended. Looking for shared interests or common experiences to discuss is much healthier, whether you are talking to new acquaintances or old friends.

Avoiding all unfamiliar topics is far from the ideal way of building social connection, however; if a subject is central to your life and represents an important element of your personality, you need to find a way to express it – otherwise your shared reality with the other person will always have an important part missing. In these cases, you can try to escape the novelty penalty with vivid storytelling that helps to put the other person in your shoes, so that they can better understand exactly why this event or piece of knowledge is so important to you and why it should matter to them. If you know that the person is a gastronome, for example, it makes sense to start out by discussing the food you ate on the trip, which should act as a bridge to their own interests and experiences.

As you move onto less familiar terrain, you must make sure that you provide enough details to avoid creating unnecessary informational gaps. Think carefully about their baseline knowledge, so that you don't patronise them – if necessary, you should ask how familiar the subject already is – and use this to gauge the elements that you need to include to ease their understanding. Your estimates are going to be imperfect, but given the research on the novelty penalty, it seems worth veering on the side of caution by repeating a few familiar facts.

In Cooney's experiments, speakers reduced the novelty penalty if they gave a more complete narrative of the videos under discussion. When they were recounting the recent scientific discoveries about crow intelligence, for example, it helped to describe the inspiration for the research, and a general overview of the main conclusions (crows are smart!), followed by more in-depth accounts of the individual findings. They ended by describing how we could train crows to pick up litter in sports stadiums – and how our understanding of crow intelligence might change the way we think about the human mind. With this level of detail, the speakers enjoyed the discussions almost as much as the discussion of the topic that was already familiar.

The importance of crafting a narrative may sound obvious to anyone who spends their life in written communication, but many of us forget to apply the same rules to our conversation. My lingering shyness has often led me to rush my descriptions of new or exciting experiences; the quicker and more concise my story, the less chance that I'll be boring the other person, I thought. This conversational approach was the equivalent of presenting a pencil sketch with all the colours missing: my pithy descriptions simply weren't giving people enough information to understand why those events had been meaningful to me, which had imposed an unnecessary barrier to connection. I was especially reluctant to disclose my emotional reactions to events, yet these represent exactly the kind of self-disclosure that helps to build connection. Cooney's research has encouraged me to be more confident in relaying what I see, think and feel, safe in the knowledge that those details will help the other person to relive the experience with me.

Besides reconsidering the ways that you tell stories, you might also remember the novelty penalty when the roles are reversed, and you are struggling to engage with someone else's experiences. In the past, your general reluctance to ask questions might have prevented you from seeking out the additional information that would allow you to close the gap in understanding. (You might have even written them off as a bore.) Once you have recognised the novelty penalty as the source of these feelings, however, you can use the occasion to show your willingness to create a shared reality and enter their world. If you encourage them to open up and immerse you in descriptions of their experience and the things that most matter to them, you may find their story to be far more engaging than you initially expected. When appropriate, you can then provide your own interpretation of what they are describing and give them accounts of your experiences that might provide additional points of connection.

Whoever we are talking to, and whatever we are talking about, we should be looking for balance – in the exchanges between partners, in the depth of the discussion and in the familiarity of the topics. This is the crux of our fifth law of connection: **In conversation, demonstrate active attention, engage in self-disclosure, and avoid the novelty penalty, to build mutual understanding and contribute to the merging of our minds.** Whether we are on a first date or meeting a lifelong friend, each sentence we speak offers a new opportunity for greater connection.

What you need to know

- Asking relevant questions and showing signs of careful listening are two of the easiest ways to build connection
- Contrary to standard advice, interruptions can improve conversational flow – provided that they are well timed and express joint understanding
- Self-disclosure is more effective than small talk at building a rapid rapport – and people are more interested in your deepest thoughts than your instincts lead you to believe
- The 'novelty penalty' means that we are often safer to chat about familiar topics than new and exciting experiences. This arises from informational gaps that make it hard for people to relate to what you are saying – but good storytelling can help you overcome this barrier

Action points

- Consider your conversational style and the types of questions that you ask of other people. Are they well-selected follow-ups that build on the points they've just raised, or do you engage in boomerasking that simply turns the conversation back to you?
- How do you express your attention to other people? Non-verbal cues – such as nodding your head or murmuring assent – are often not enough to convince someone that you really care; it's far better to verbalise your interest by, for example, paraphrasing what they have just said
- Attempt the fast friends procedure with one willing friend or acquaintance – and try to incorporate more self-disclosure in your everyday conversation
- Be compassionate when you find that you are bored of a conversation and ask yourself what you can do to help the other person to express the main point of interest and the reasons that the subject matters so much to them

CHAPTER 6

EXPRESSING APPRECIATION

Now that we have explored the art of conversation, our next law of connection concerns the ways that we show our appreciation of others. Of all our social interactions, praise and gratitude should be the easiest to express and the most pleasant to receive. Yet Western culture has long warned us to be wary of honeyed words.

Consider one of Aesop's most famous fables, in which a fox is hungrily wandering the forest looking for a bite to eat, when he finds a crow sitting on a branch with a piece of cheese in her beak. The crow is initially wary and turns away, but the fox love-bombs her with a series of compliments, admiring the lustre of her feathers and the handsome shape of her wings. 'Such a wonderful bird should have a very lovely voice, since everything else about her is so perfect,' the fox concludes. 'Could she sing just one song, I know I should hail her Queen of Birds.' Charmed, the crow starts to caw, dropping the piece of cheese straight into the fox's open mouth. The moral is clear – praise is a tool that can be used to prey on the vain and gullible.[1] As Jean de la Fontaine noted in his version of the fable, 'Flatterers thrive on fools' credulity/The lesson's worth a cheese, don't you agree?'[2]

Dante provided a more gruesome warning. In the *Divine Comedy*, he places flatterers in the second pit of the eighth circle of hell, where they are immersed in a river of diarrhoea – the external filth providing a fitting punishment for all the insincere words that had fallen from the sinners' mouths during their time on Earth. The brown-nosers – who include one of Dante's political rivals – are so deeply buried in the excrement that they are almost impossible to recognise.[3] Dante was, of course, following a biblical proverb: 'A flattering mouth works ruin.'

Other examples include the sycophantic Mr Collins in *Pride and Prejudice*, Uriah Heep in *David Copperfield* and Gríma Wormtongue in *The Lord of the Rings*. Thanks, perhaps, to the lingering influence of these tales, many of us are still extremely wary of expressing our admiration for others, in case we seem awkwardly obsequious. That is a great shame, since current psychological research shows that genuine expressions of praise or gratitude can have huge benefits for both giver and receiver. Exchanging kind words not only motivates us to behave better; it also calms our stress responses, increases our resilience to change and reduces our risk of mental illness. To do so, however, we must overcome our fear of being misinterpreted.

THE PRICE OF KIND WORDS

Let's start by exploring some of the benefits of praise for cooperation and collaboration. From a purely scientific viewpoint, there can be no doubt that praise and appreciation are priceless motivational tools. Many studies on this theme were conducted in the workplace, with the surprising revelation that verbal commendations can be more valuable than financial bonuses. The organisational psychologist Liad Bareket-Bojmel tested out various bonus schemes on the technicians at one of Intel's factories in Israel. On varying weeks, they were offered either a cash prize of 100 shekels (the equivalent of US$25 at the time), a pizza voucher of the same value or a thank-you note from a senior manager. The employees' productivity was recorded before the trial started and then every eight days; they would receive the reward if they exceeded the average performance at the beginning of the study.

As you might hope, all the rewards increased the employees' commitment, measured by the number of semiconductors the factory produced, and the simple verbal recognition did just as well as the cash prize at inspiring greater effort. The wording of the note was not even particularly well chosen. It simply stated, rather impersonally: 'Thank you for your hard work and great achievements in yesterday's shift, I appreciate your effort very much.' Yet the small sign of gratitude must have meant something to the people who received it. Tellingly, the employees continued to work hard on the days after receiving the note, whereas the people who had been given the cash reward showed a noticeable drop in productivity when the bonuses had been removed – perhaps because they now started to resent the fact that they were no longer being compensated for any extra effort.[4]

Praise seems to be particularly powerful for administrative tasks that offer little intrinsic interest, such as data entry. In such cases, a simple thank-you note not only inspires those who have received the recognition; it motivates the people around them to work harder too.[5] Public praise was also considerably more effective at encouraging Dutch employees to adopt energy-saving policies than cash bonuses.[6]

Unfortunately, many employers neglect these benefits. One poll of 2,000 US citizens found that 63 per cent of people felt underappreciated most days, while 59 per cent said they had never had an employer who truly recognised their work.[7] Another study, of more than 200,000 people, found that a general lack of appreciation was the primary reason for employees' resignations.[8]

The equivalence of fine words and monetary recompense can be seen in the brain's responses to both kinds of reward. Japanese researchers invited visitors into their lab and asked them to play a simple game of chance for small cash prizes as they lay in an fMRI scanner, which allowed the scientists to identify which brain regions respond to financial incentives. Afterwards, they answered personality questionnaires, wrote a short essay about a political or social issue of their choice, recorded a one-minute video introducing themselves and posed for photos – all of which, they were told, would be evaluated by eight strangers. In reality, the researchers made up the responses themselves, some of which were more flattering than others. A few days later, the participants were invited back into the lab and were presented with the ostensible feedback as the fMRI scanner registered their brains' reactions to the comments. As the behavioural experiments had suggested, the brain activity in response to the cash prizes and the social praise was strikingly similar, with heightened activity in areas such as the striatum, a region deep in the brain that is associated with elemental feelings of pleasure.[9]

Providing people with that sense of reward can be highly persuasive in many different contexts. Indeed, according to a study by Naomi Grant at Mount Royal University in Canada, even the most casual of compliments can encourage strangers to be more cooperative. Grant first invited participants to take part in a study of 'impression formation'. As they filled in a rather dull questionnaire, an actor – who posed as an introductory psychology student – struck up a conversation that involved casually complimenting the participant's clothing. Following the chat, the actor mentioned that they were handing out leaflets about a careers event, and asked the participant whether they would like to take a handful to distribute themselves. The effects of the flattery were dramatic. Seventy-nine per cent of these participants offered to help with the event publicity, compared with only 46 per cent of participants in a control group of people who had not received the compliment during the chat.[10] Grant's later research suggested that this comes from a sense of reciprocity: if you scratch my back, I'll scratch yours.[11] In this way, compliments resemble the 'social grooming' behaviours of our early primate ancestors. Monkeys and apes may solidify their relationships by picking ticks out of each other's fur and repay the efforts with support during conflicts. We do the same with the exchange of good words and deeds.[12]

Praise can be persuasive even if the flatterer's ulterior motive is on full display. Psychologists at the Hong Kong University of Science and Technology, for example, gave each participant an invitation to visit a new department store, where they would be asked to evaluate its services. 'We are contacting you directly because we know that you are a fashionable and stylish person. Your dress sense is not only classy but also chic. As someone with exceptional taste in clothes, you will enjoy the designs featured in our new collection, featuring "must-haves" for the coming season.' When questioned afterwards, the recipients of the invitation reported being perfectly aware that the flattery was being used to encourage their participation. For many people, however, the implicit feeling of being appreciated lingered, and predicted a greater desire to shop at the store a few days later.[13]

R.E.S.P.E.C.T.

If you are of a Machiavellian disposition, you may be interested in using flattery to exert influence over others: there is, perhaps, a hint of truth in Aesop's fable of the crow and the fox. For the purposes of this book, however, I am less interested in the power of praise to persuade than its benefits for social connection. A simple compliment to a stranger might be a good opener for a meaningful conversation or even prompt the start of a friendship. Among established social ties, sincere appreciation leads people to feel better understood and provides the sense of security that is necessary for open communication. At the very least, a compliment or word of thanks should provide an instant mood boost that brightens an otherwise dull day.[14]

Unfortunately, many of us are stingy with our praise. Think about your own life. Have you ever admired someone's sense of style, but failed to mention what you think? Or perhaps you've attended a talk but never let the speaker know how much you enjoyed their speech? As a journalist, I know how much it means to receive a quick note of thanks for an article I've written – but I rarely reach out to offer that gift to the people whom I admire. We're only a little more generous with the people in our direct social circle. When asked to consider their behaviour towards one of their closest friends or family members, for instance, the average person estimates biting back around 36 per cent of the compliments that come to mind. This did not seem to be influenced by how close they felt to that person.

Even if we hold someone very dear, we regularly hold ourselves back from telling them so.[15]

Erica Boothby and Vanessa Bohns at Cornell University were among the first scientists to uncover the psychological biases that prevent us from sharing our goodwill. In the first study, participants were asked to walk to a designated area of the campus – such as a dining hall or lobby – and compliment the fourth person they saw on an item of their clothing. Just beforehand, they completed a survey asking how they expected the other person to react to their words – rating, for example, how pleased and flattered they assumed the other person would feel on a scale of 1 (not at all) to 7 (very much). After giving the compliment, they handed the person a sealed envelope that contained a similar set of questions testing the way they actually felt. The original participant then returned to the lab with the completed survey. As predicted, the compliment-givers consistently underestimated how good the compliment-receiver would feel. They just didn't realise how much their kind words would brighten the other's mood.

Echoing the original liking gap, further studies showed that people also overestimated the costs of the conversation. They predicted that the person receiving the compliment would find the interaction far more awkward than it actually was, and that they'd come across as a nuisance. In reality, most people were delighted to hear the compliment and felt little discomfort at the exchange.

Bohns and Boothby's final study demonstrated that these fears arose from participants' general under-confidence and anxiety about their overall social competence. Participants thought that they were uniquely bad at giving compliments, leading them to be overly pessimistic about the ways their words would be received. When considering how other people would perform a similar task, however, their predictions tended to be much closer to the mark.[16] As we are seeing time and time again, we have greater potential for social connection than we realise; we must simply give ourselves more opportunity to use it.

Like the original liking gap, this seems to be an incredibly robust phenomenon, that has been replicated in a large study by Nicholas Epley at the University of Chicago and Xuan Zhao at Stanford University. Whereas Bohns and Boothby's samples consisted primarily of students, they recruited people from the local community, to demonstrate that the results generalise to people of diverse ages and backgrounds. And rather than focusing on exchanges between random strangers, they asked their participants to compliment someone who was already within their social circle.

To do so, Epley and Zhao set up a table in a local park, with a poster advertising an Interpersonal Relationship Study. This allowed them to recruit pairs of acquaintances who had been taking a walk together. After a slight preamble, the researchers asked one person from each pair to write down three compliments to tell the other person: 'positive things you have noticed but have not, for whatever reason, had a chance to compliment your partner on yet'. They then completed a questionnaire about how pleased they thought their friend would be to hear what they had to say. Their compliments were then passed on to the other person, along with a questionnaire asking them to rate how they felt about these words of praise. Once again, the participants were far too pessimistic in their predictions of their friends' reactions, which were much more warmly received than anticipated. They assumed they wouldn't be able to find the right words to adequately express their admiration. In their acquaintances' eyes, however, the warm feelings came across loud and clear.[17]

Can we ever take our compliments too far? Too much sugar can ruin the tastiest of cakes, after all, and it stands to reason that our acquaintances may begin to find our honeyed words a bit cloying if they are offered too frequently. Yet further work by Zhao and Epley suggests that this reaction is far less likely to happen than we believe. In this experiment, the researchers again recruited pairs of participants who already knew each other. One member of each pair was asked to note down five separate compliments for their acquaintance. The researchers then delivered these compliments to the recipient, one by one, over the following week, and asked them to fill out surveys describing how they felt with each new piece of praise. Their pleasure did not decline significantly over the week; nor were the words of praise perceived as being any less genuine on the last day compared to the first; indeed, each compliment brought some new joy into their life.[18]

Such a consistent mood boost is unusual in psychology. Very often we become habituated to particular pleasures if they are repeated too often and need to look for other sources of enjoyment, a process known as 'adaptation', but there was no evidence that this applies to our compliments. This is not so surprising when you consider that social connection is a fundamental human need, like food or water, that must be constantly replenished, rather than a luxury that quickly loses its shine. By regularly offering others your kind words, you are actively helping people to satiate their social hunger.

These conclusions could not be further from most historical etiquette advice. Consider the words of the eighteenth-century writer Samuel Johnson. Like William Hazlitt, Johnson was acutely interested in the art of conversation, and argued that flattery was best applied rarely. 'Praise, like gold and diamonds, owes its value only to its scarcity,' he wrote in an essay for *The Rambler*. 'It becomes cheap as it becomes vulgar, and will no longer raise expectation or animate enterprise. It is therefore not only necessary, that wickedness, even when it is not safe to censure it, be denied applause, but that goodness be commended only in proportion to its degree; and that the garlands, due to the great benefactors of mankind, be not suffered to fade upon the brow of him who can boast only petty services and easy virtues.'[19]

Johnson was correct that compliments can act as precious gifts – but there is no reason to parcel out our praise in the ways that he suggested. We can all be more generous with our words of admiration without running our stock dry or diminishing their impact on the people who receive them.

THE GRATITUDE GAP

If you are already feeling inspired to express more appreciation to the people around you, you might begin by sharing your feelings of gratitude. Simply send an email or text message, right now, to someone who has recently helped you. You'll both feel better for it, and – as we shall soon discover – this simple act may boost your physiological resilience to stress.

All too often, we assume that others know how much their actions mean to us, but our sentiments are rarely as obvious as we believe, and so it is worth saying them explicitly. MBA students at the University of Chicago, for example, were encouraged to write a thank-you letter to someone who had touched their life in a meaningful way. The letter, they were told, should describe what the person had done for them, how the action had affected them and why they felt grateful for their behaviour. Afterwards, the students answered a range of questionnaires measuring their perceived competence at expressing their feelings of gratitude effectively, how the other person would feel to receive the letter, and whether the other person was already aware of their sentiments. 'We're also interested in the extent to which you feel like the person you sent this letter to already knows the things you wrote down,' they were told. 'That is, how surprised do you think they will be to learn about the specific reasons for why you feel grateful to them?' The participants could answer on a scale ranging from 0 (not at all surprised) to 10 (extremely surprised). Nicholas Epley and his colleague Amit Kumar then passed on the notes to the recipients and asked them to complete a questionnaire detailing their feelings about the letters.

Echoing the studies of people giving compliments, most of the MBA students sending the thank-you notes underestimated how much their kind sentiments would be appreciated by the recipients. Equally importantly, however, the participants also misjudged how much the recipients already knew about the way they felt. To the person sending the letter, egocentrically immersed in their own point of view, it seemed obvious that the recipient would realise how much their actions had meant to them, but this was not the case. Most of the recipients were surprised to hear that the sender felt this way.[20]

Rather wonderfully, the benefits of expressing appreciation are not limited to the person receiving the kind words. In almost all the studies mentioned, the people offering the kind words ended up feeling considerably happier after the experience. And if we practise this regularly, it can boost our mental health and resilience. People who wrote three sincere thank-you letters over three weeks reported higher levels of life satisfaction and lower levels of depressive symptoms, for example.[21]

By soothing the troubled mind, expressing and receiving appreciation can also calm the body's physiological stress response. Yumeng Gu and Christopher Oveis at the University of California, San Diego asked pairs of university roommates to take part in a game inspired by the TV programme *Shark Tank* (known as *Dragons' Den* in the UK) in which aspiring entrepreneurs propose a new business plan to a panel of expert investors. The roommates had only six minutes to prepare their pitch, and the stakes were relatively high: the best pair would receive $200 of prize money – a considerable sum for a student – meaning that the students would feel under considerable pressure to perform at their best.

Before the students entered the 'shark tank', the researchers asked one member of each pair to think of a time when their partner had improved their life in some way, and to express that to him or her in a short conversation. They were then hooked up to electrodes measuring their heart activity and blood pressure cuffs, which allowed the researchers to see how the participants' bodies responded to the challenge of planning and pitching their business proposal. As a comparison, a separate group of roommates went through an almost identical procedure, but without the gratitude exercise. Most students showed signs of a stress response during the task, but those who had either said or heard the words of appreciation revealed a healthier reaction. Both members of these pairs showed better circulation with easier blood flow across the body, so there was less strain on the heart. This was a stark contrast to the participants in the control group, who showed characteristics of the fight-or-flight response, where the body reduces blood flow to its peripheries as a means of limiting the damage from a potential injury.[22]

The fight-or-flight response is generally associated with worse cognitive performance, and, if we experience it regularly, poorer overall health. We cannot always avoid stress in our lives, but any strategy to reduce the physiological reaction should reduce its long-term consequences. Gu and Oveis's study shows that simply sharing moments of gratitude can be a productive way of buffering the effects – and it does not seem to matter whether we are expressing or receiving the words of appreciation. Why should gratitude bring such benefits? The simplest explanation is that the extra sense of social connection unconsciously increased the roommates' sense of the resources at their disposal; whether they were expressing or receiving gratitude, the participants knew they had each other's backs and that they could draw on their combined knowledge and mutual support to meet the demands of the situation. As a result, they felt that they were under less of a threat – and their bodies responded accordingly.

Whatever the reasons, Gu and Oveis's findings are worth remembering whenever you or your acquaintances are facing personal or professional challenges. In stressful circumstances, it is easy to overlook social niceties like compliment-giving and expressions of gratitude – but it is precisely these behaviours that will boost your resilience and help you to cope with the situation in the most productive way possible. Along these lines, one study from the University of North Carolina asked couples to have regular twenty- to thirty-minute conversations in which they expressed gratitude to each other for their support. 'Your partner's positive gesture may be something that happened before but continues to make you grateful, or something going on now,' they were told. 'Some examples would be helping to solve a problem, surprising you with a gift, taking time to listen to a concern, spending time doing something he or she would not typically do, or similar things.' They were asked to have four to six of these conversations over a month. Those in the control group were asked to engage in greater self-disclosure more generally. It's a high bar for comparison, since self-disclosure should itself strengthen our sense of connection.

The benefits, as you might expect, depended on the couple's commitment to the intervention. When people failed to practise this regularly, their wellbeing was not better than the control group's. Those who took the task seriously and frequently engaged in open and responsive discussions, however, experienced an increase in psychological resilience, greater capacity to adapt to changes in their lives and higher overall relationship satisfaction.[23]

If you read the wellbeing pages of newspapers or watch the health segments of daytime talk shows, you will have almost certainly heard of 'gratitude journalling' as a strategy to improve emotional wellbeing (Oprah Winfrey has been one of its biggest champions). Sadly, far fewer people have so far adopted the practice of sending thank-you notes – despite all this evidence that it can bring many more benefits, not just to our immediate wellbeing but to the people around us and the quality of our relationships. This preference is not surprising: our erroneous intuitions mean that we are far more likely to seek a solitary activity such as journalling than a form of self-disclosure that triggers our fear of social judgement.[24] But the scientific research shows that those worries are unfounded. If we are already thinking something kind about someone, then there is rarely a logical reason to keep those thoughts to ourselves.

THE WITNESSING EFFECT

As we show our appreciation for another person, we will have to decide whether to do so in private or whether to reveal our admiration openly. The first option may work best if we are dealing with someone who is particularly shy and dislikes being the centre of attention. You may also worry about triggering jealousy and resentment in the people who are not receiving your praise.

There are, however, some serious benefits to letting our sincere appreciation be known more widely, which may persuade us to be a bit more open about our admiration. In a series of studies on gratitude, Sara Algoe at the University of North Carolina and colleagues found that people who had observed a display of gratitude became more cooperative with the person giving thanks, and they felt freer to open up about their deeper thoughts and feelings. Algoe and her colleagues call this phenomenon the 'witnessing effect'.

In the first experiment, the participants were asked to read a movie review and highlight the most important sentences within the document – feedback that, they believed, would be given back to the original writer. Before they completed the task themselves, they looked at a previous example – ostensibly from another participant – which contained a comment bubble from the author thanking the person for having caught some typos. Despite there being no instructions to do so, the real participants were now considerably more likely to look out for typos in their own task. This was not the case if they saw a previous document that contained typo corrections in the tracked changes, or a document with a note that instead said 'Congratulations on finishing the editing'. They had to witness the explicit expression of gratitude to be inspired to do good themselves.

Further experiments revealed that witnessing a small display of gratitude by a stranger promotes a greater desire to form a friendship with that person. The participants watched a video of someone describing their participation in a recent running race. In some videos the runner mentioned how their romantic partner had cheered them on through the race, and emphasised how much they'd appreciated the gesture; in others, they talked about pride and feelings of accomplishment. After watching the videos, the participants were allowed to write a short letter to the person in the video describing a recent positive experience, and they tended to share more personal information, about their deeper thoughts and feelings, if they had been shown the video including gratitude to their partner. Since self-disclosure is so important for connection, this was taken to be a sign that the display of gratitude had primed them to build a close bond. Sure enough, these participants were also more likely to agree with the statements 'I would enjoy meeting the person in the video' and 'I can see myself being friends with the person in the video'.[25]

We could all benefit from the witnessing effect. By showing appreciation for others, we send out an important signal about the kind of person we are. It shows that we are responsive and supportive to the people within our social network – valuable qualities in any relationship.

FLATTERY WITHOUT SYCOPHANCY

An increased capacity to feel and express our appreciation for others might have helped drive the social revolution in our ancestral past. As we started to live in bigger groups, we'd have needed new psychological mechanisms to strengthen relationships and ensure cooperation. We've seen a few of these already – such as music, dance and synchronised activities that can act as social glue. And the expression of gratitude may be another.

Sara Algoe, who conducted the research on the witnessing effect, proposes that it does this through three main mechanisms, which she describes in her 'find–remind–bind' theory:

Find: gratitude helps us to identify a stranger who is worth cultivating as a friend.

Remind: gratitude draws our attention to the people who are already in our social network that support and help us, so that we do not take them for granted.

Bind: gratitude motivates us to make a greater investment in those relationships.

We've already explored some examples of the ways that compliments fulfil these functions, and we'll now discover how science can offer some additional guidelines to maximise the benefits of these gestures.[26]

Together, they constitute our sixth law of connection: **Praise people generously, but be highly specific in your words of appreciation.** This comes back to the idea of the shared reality – people want to know that you are paying a lot of attention to them, since that will help you to understand how they are thinking and feeling. While people may graciously receive the odd platitude as a means of generic social grooming, vague words do not demonstrate much interest in the unique qualities this person has to offer, and they can easily be faked. Clichéd compliments are not going to bring nearly as much joy as a sign that you've noticed and appreciated some unique facet of the person that no one else had observed.

The same rule goes for gratitude. There are two ways we can express our thanks: by describing the benefits of an action for us, or by praising the personal qualities that contributed to the act of generosity. You should aim for the latter. Imagine, for example, that your partner has just given you a rare book by one of your favourite authors. You could say 'thank you, this looks awesome, I can't wait to read it' – which is a polite enough statement that demonstrates how their behaviour has benefited you. But it would be even more powerful if you go on to observe the thought that has gone into their choice, and acknowledge how well they must know you to have spotted the book and bought it. Similarly, if your friend has just picked you up from the hospital, you might say 'thank you, that saved me a long bus journey'. But your words may mean even more if you reflected on what that says about them, as a person: 'I really appreciate how you always go out of your way to help me.' Research by Sara Algoe suggests that these small changes can make all the difference in the way your words are received.[27]

We can see this in a study of couples who had been encouraged to express more gratitude to each other for one month. The intervention successfully improved the average participant's wellbeing and their relationship satisfaction, but the effects depended on the complimentors' 'responsiveness': how much their conversations left their partner feeling understood and validated.[28] Simply saying 'you look nice' or exclaiming 'good job!' isn't as likely to do this as is careful attention to the personal qualities that you appreciate. For similar reasons, we should use hyperbole sparingly. No one wants to be compared to a celebrity to whom they have no honest resemblance.

We should be especially wary of letting our implicit prejudices shape our compliments. Lazy clichés based on 'positive stereotypes' that may superficially seem flattering but reflect an implicit bias are more likely to offend than arouse pleasure. If you only ever praise a gay man for his fashion taste, then you're not appraising the individual as much as the category that he belongs to – and it's unlikely to increase feelings of social connection as a result.[29] (That's unless, of course, you are giving genuinely well-considered compliments that accurately reflect the individual's achievements and values, rather than your prejudiced views of them.)

We should also be aware of someone's confidence and adapt our compliments to their view of themselves. People with low self-esteem have a tendency to dismiss compliments and may even feel anxious or ashamed at hearing words that contradict their beliefs.[30] One study even found that people with low self-esteem show a greater sense of commitment to their relationship when their partners had expressed a *poor* opinion of them that matched their beliefs, compared with when their partner had described them in a more flattering light.[31] Remember that, when forming a shared reality with someone, we want people to verify our view of the world – and if someone has serious doubts about their worth, they might suspect that the other person's praise is dishonest, or they may fear that the compliment is based on a false assumption, which will soon be uncovered. Or they assume that the other person was 'just being nice', without really meaning what they said. Both ways of thinking will weaken the sense of connection binding the two people.

In such cases, it will be especially important to describe in high detail why we appreciate the quality in question and to explain its importance for the overall relationship. Since people with lower self-esteem tend to assume our feelings are fleeting, we might also make more of an effort to emphasise the continuous nature of our appreciation. Telling someone that you *always* admire their creativity whenever you have a problem, for instance, and describing a few times in which they'd helped you find your way out of a rut, may be more likely to activate their brain's reward centres than simply thanking them for their support on that one occasion.

Encouraging people with low self-esteem to think about others' compliments in these ways increases their sense of security in the relationship. Research from the University of Waterloo in Canada, for instance, found that participants were considerably more likely to endorse statements such as 'My partner loves and accepts me unconditionally' and 'I am confident my partner will always want to stay in our relationship' after they had been asked to think more carefully about the reasons why their partner admired them, following a compliment.[32] Our words alone cannot eliminate the person's deeply held insecurities, but we can help them to see themselves through our eyes, which should help them to trust that our appreciation is genuine and lasting.

Praise in the workplace must consider the organisational hierarchy and the power dynamics at play. If we are of a higher social status – a boss dealing with a junior member of staff, say – we might be tempted to qualify our compliments as a way of maintaining the hierarchy. (A manager might tell a trainee, for example, that 'Your ideas were good . . . for an intern.') Backhanded compliments of this kind tend to reduce the motivation of the person you were trying to praise. Worse still, they signal our heightened concern about our image, which makes us seem less likeable and damages our credibility.[33] If we have something nice to say to someone, we should simply express the sentiment without adding any bitter notes.

Finally, we should be sure not to overlook expressions of gratitude to our weak ties – those people who we barely know at all, but who may one day join our closer social network. In one striking experiment, researchers asked university students to give quick feedback on a high school student's writing. A week later, the mentors were casually asked if they would be willing to meet up with the student they had helped, and if so, to name the best way to contact them. In some cases, they had been given a thank-you note before the request; for others, there was no personal acknowledgement of the help they'd given. The effects were huge: mentors were around 63 per cent more likely to pass on the information for a later meet-up if the mentee had followed up with that quick expression of gratitude. The expression of gratitude had acted exactly as the find–remind–bind theory would have predicted, priming the participants for social connection.[34]

Many of us are already in regular contact with truly remarkable people: we're simply too shy about showing our appreciation. Once we overcome this barrier, and apply the sixth law of connection, we will see our social networks growing in size, strength and happiness.

What you need to know

- People often value words of appreciation and praise as much as physical gifts, and in the workplace they can be a serious source of motivation and a driver of productivity
- We don't praise people nearly as much as we should, due to concerns about our ability to phrase compliments correctly. We also erroneously believe that others already know how we feel
- When we express appreciation to others, it bolsters our feelings of social support. As a result, both parties see a considerable boost in their wellbeing. Showing gratitude to others can even buffer our stress response in challenging situations
- Public praise not only benefits the direct recipient, it can also signal your desire to form strong social connections, leading you to stronger bonds with the others in your group. This phenomenon is known as the witnessing effect

Action points

- Send a message to someone whom you have not seen in a while. Let them know why they are important to you and remind them of a time that you appreciated their company. You may be surprised by how effectively it rekindles your connection
- Avoid platitudes and lazy clichés. To build a sense of connection, your expressions of appreciation must be grounded in your shared reality with the other person. They should demonstrate careful attention to the person's unique qualities and their deeds, with specific details about the reasons you appreciate them
- If someone instantly dismisses your compliment, it may be that the words you used challenged a negative view of themselves, provoking anxiety rather than pleasure. You can overcome this by emphasising the duration of your positive feelings and by more carefully expressing the reasons this quality is so important for your relationship

PART 2

MAINTAINING CONNECTIONS

CHAPTER 7

TRUTH, LIES AND SECRETS

The works of the nineteenth-century German philosopher Arthur Schopenhauer may seem like an unlikely place to look for the secrets of social connection. Having developed a deep melancholy in childhood, Schopenhauer is famous for his bleak outlook on life and his disagreeable personality. He was, by all accounts, intolerant and impatient, and had an extremely high opinion of his own talents that did not always match his achievements. Schopenhauer's jealousy would sometimes drive him to absurd measures. While teaching in Berlin, he scheduled his lectures at the same time as those of his rival, the philosopher Georg Wilhelm Friedrich Hegel, a man he described as a 'clumsy charlatan'.[1] Schopenhauer may have held little regard for Hegel, but many other intellectuals considered him to be the greatest mind in Europe, and students flocked to hear him speak, while Schopenhauer spoke to a near-empty lecture theatre.[2] After once failing to win an important prize for his work, he responded by writing a tract denouncing the incompetence of the judges.[3]

As we have also seen with Isaac Newton, however, a truculent personality does not preclude deep and influential friendships. Schopenhauer's social circle included the merchant Jean Anthime Grégoire de Blésimaire, whom he first met in childhood; the polymath Johann Wolfgang von Goethe; the scholar Karl Witte and the American business magnate William Backhouse Astor.[4] And his writing reveals a shrewd awareness of the importance of connection and the psychological barriers that prevent us from forming close bonds – as seen in the 'porcupine's dilemma'.

'A number of porcupines huddled together for warmth on a cold day in winter; but, as they began to prick one another with their quills, they were obliged to disperse,' Schopenhauer wrote in *Parerga and Paralipomena*, a collection of philosophical essays. 'However, the cold drove them together again, when just the same thing happened. At last, after many turns of huddling and dispersing, they discovered that they would be best off by remaining at a little distance from one another.'

Humans behave much the same, Schopenhauer suggested. The closer we try to get to other people, the more likely we are to be hurt by their behaviour. This leads us to set ourselves at a 'moderate distance' and shield ourselves through mutually polite manners that prevent either side from getting hurt. 'By this arrangement the mutual need of warmth is only very moderately satisfied; but then people do not get pricked. A man who has some heat in himself prefers to remain outside, where he will neither prick other people nor get pricked himself.'[5]

We have now learned many ways to build stronger relationships: overcoming the liking gap; avoiding illusions of understanding; mastering the art of conversation; and successfully expressing appreciation. These are the foundations of connection. But close relationships with other people will inevitably involve discomfort and conflict, and in Part 2 we shall discover how to navigate these difficulties. Our intuitive solutions to these challenges can often increase our distance from others – but it needn't be this way. According to the latest social psychology, there are many ways to overcome the porcupine's dilemma.

We'll start with a discussion of truth, lies and secrets. Social connection is established on the premise that we are both operating honestly – otherwise there is no way that we can create a shared reality. As we begin to get to know each other more deeply, however, there are bound to be things that we feel afraid to disclose, for fear that it will muddy another's image of us, or, worse still, cause the other person emotional pain. Is it better to 'let sleeping dogs lie' or should we come clean about the elements of our past that we'd rather forget? Are we ever justified in telling little white lies?

I don't think it will be too much of a spoiler to say that the answer, in almost every possible scenario, is to act a little more bravely than we feel. This will be our seventh law of connection: **Be open about your vulnerabilities, and value honesty over kindness (but practise both, if possible).**

THE BURDEN OF SECRETS

Like most members of the LGBTQ+ community, I knew from a young age what it means to keep an important part of myself hidden from others.

While I can't remember when I first knew I was gay, I can recall knowing, for the first time, how hard it would be to share my sexuality with others. I was around ten years old and my family were staying with some elderly relatives for a few days. It was the evening, and as my brother and I sat watching a sitcom on TV, the adults' conversation turned to the topic of a former lodger. He'd been kicked out, my great-uncle declared, for having had men in his room. His tone, and my parents' silence, were all the evidence I needed to know that my infatuations with male stars should remain a forbidden subject.

At first, the secret didn't feel too burdensome, but it was lonely knowing something about myself that I couldn't share with anyone else. By being constantly alert to any potential display of my sexuality, I couldn't help but hold myself back in conversation; the spectre of being discovered left very little room for self-disclosure of any kind. By the time I came out to my friends as a teenager, I knew this was not the life I wanted to live. It was the first time I had truly understood the truth in André Gide's assertion – famously repeated by Kurt Cobain – that 'it is better to be hated for what you are than to be loved for what you are not'.[6]

Psychological research confirms that keeping secrets can be a huge source of stress, with serious consequences for our health and our relationships. And it is something that almost everyone experiences. According to a detailed survey by Michael Slepian and colleagues at Columbia University, the average person held thirteen secrets, five of which they have not told to a single person. Fewer than 3 per cent – 3 per cent! – of people reported not having any secrets at the time of answering the questionnaire.[7] As you might expect, some of the most common secrets concerned sexual desires and romantic infidelities, but the participants also reported unspoken ambitions, self-harm, cheating in the workplace, personal traumas and unpopular political beliefs.[8]

The most obvious source of stress would be the sheer effort of concealment. Someone who is hiding their sexuality will naturally feel uncomfortable discussing dating; someone who is suffering silently from an eating disorder may struggle when a conversation turns to diet and food; and someone who has lied about what university they attended will break into a cold sweat whenever education is mentioned. But Slepian's research suggests that this stress lingers even in our downtime.[9] In much the same way that our tongues can't seem to help prodding at a tooth that hurts, when we have a serious secret, our minds constantly return to our fears of being exposed.[10]

The mental strain of suppressing information can dim our intellectual performance. Researchers at the University of California, Berkeley and Cornell University asked straight participants to have ten-minute conversations without revealing any clues about their sexual orientation. If they spoke about their dating preferences, for example, they had to describe their partner in gender-neutral terms. Clearly this only weakly imitated the experience of someone who, fearing social stigma, decides to conceal an important part of their identity. There was no real risk of hostility if they slipped up, after all, and a one-off ten-minute interaction is hardly comparable to a lifetime of suppression. Nevertheless, the straight participants found the mental vigilance and cognitive gymnastics to be significantly fatiguing, reducing their scores on subsequent IQ-style tests of non-verbal reasoning.

We often experience this strain as a sense of physical burden;

when a secret is meaningful and central to our sense of self, it almost literally weighs us down. During one series of eye-catching studies, Slepian's team asked participants to think about a secret that they had been carrying, without revealing the specific details. Some were told to think about something trivial, while others were asked to think of something more profound that they had kept hidden. The participants then had to estimate the steepness of a hill shown face-on. Previous research had shown that to people carrying extra weight – such as a heavy backpack – hills appear to be steeper than they are. And Slepian noted an identical response in those churning over the big secret: they overestimated the gradient of the slope by 40 per cent compared to those who were thinking about something more trivial.

For a second experiment, the researchers looked at distance perception. Once again, previous studies had shown that people tend to overestimate distances when they are carrying a heavy backpack – and Slepian suspected that ruminating on a big secret would have the same effect. To investigate, his team asked participants to throw a beanbag at a target 2.6 metres away. As expected, those who had just recalled a serious secret tended to be far less accurate, overshooting by around 16 centimetres compared to those who thought of something less important. The scientists next recruited people who had cheated on their partners, and primed them to think about their infidelity before asking them about the amount of effort it would take to carry the groceries upstairs, walk a dog or help someone move house. They found that the more someone reported ruminating on their behaviour, the higher their estimates for the physical exertion involved in these tasks.

Finally, the researchers invited gay participants into the lab to take part in a study on impression formation, in which they answered some questions while being filmed. Some were told not to reveal their sexuality, while others were told that they could be as open as they liked. Before the session ended, one of the researchers casually asked whether the participant could do them a favour by helping to move some stacks of books that were close by. The participants who had been asked to conceal this essential part of their identity – and were therefore feeling the mental burden of keeping the secret – moved around half as many stacks.[11]

Compounding the stress of concealment, secrets reduce our perception of available social support. When people think of some personal information that they are hiding, they score much higher on questionnaires measuring isolation, disconnection and loneliness.[12] We face a constant conflict between our desire for connection and the fear of exposure – an internal struggle that is incredibly tiring and may contribute to the feelings of mental and physical exhaustion. As you might predict, the combination of stress and isolation takes its toll on our long-term wellbeing. The more serious someone believes their secret to be, and the more it preoccupies their mind with rumination and doubt, the worse they fare on measures of life satisfaction and physical health.[13]

In the right circumstances, we can release this strain by opening up and telling the truth. Extending his studies on the embodied feelings of burden, Slepian asked participants not only to remember a secret, but also then to reveal the specific details to the researchers. He found that those who managed to articulate the thing they'd been hiding no longer exaggerated the slant of a hill, or the distance to a target. Indeed, they performed very similarly to people who were not primed to think of secrets at all.[14] The sense of physical burden, in other words, seemed to have vanished with the self-disclosure.

When and how we choose to reveal our secrets will be a deeply personal matter; we should never feel under any obligation to share information before we are ready. Recent psychological research does, however, give us some good reasons to be a bit braver when the right time comes, with evidence that people are often more supportive about the things that cause us shame or embarrassment than we might imagine. As we shall now discover, the act of sharing our vulnerabilities is often taken as a sign of authenticity and courage that will strengthen our bonds.

THE BEAUTIFUL MESS EFFECT

Since her death in 1997, Princess Diana's life and personality have been the source of endless debate, but even her harshest critics would admit that she had an incredible capacity to connect with people. And that popularity has barely waned over the years since. In 2022, a YouGov survey found that she was still better liked than her ex-husband, Charles, who had just ascended to the throne.[15]

The public admiration seems to have arisen because, rather than in spite, of her flaws. In her controversial BBC *Panorama* interview in 1995, for example, she discussed her husband's infidelities, but also her mental health struggles and her love affairs. Some of these were 'open secrets', discussed in the press, but Diana had never before spoken about them in public and on the record. It was an extraordinarily candid conversation for such a prominent figure in the mid-1990s. Many of Diana's critics believed that she had provided her own character assassination, with one (the estranged husband of Camilla Parker Bowles, now Queen Consort) claiming that she had proven herself to be 'loopy, pretty half-witted, and possibly ought to be locked up'. And they expected the public to agree.

Such predictions could not have been more wrong. Diana's popularity soared in the days after the interview, with the *Daily Mirror* reporting that an astonishing 92 per cent of the public supported her appearance on the programme. A few weeks later, the *Sunday Times* ran a survey showing that 70 per cent of the population believed that Diana should be given an official role as a goodwill ambassador abroad.[16]

In the decades since, we may have become accustomed to confessional interviews from celebrities, but we don't seem to have absorbed this in our personal lives. We overestimate how harshly we will be judged when we reveal a weakness or failure, and underestimate how much people will appreciate our honesty or courage. In general, people's perceptions of vulnerability are far more positive than we imagine – a phenomenon sometimes known as the 'beautiful mess effect'.

Some of the first evidence for this phenomenon comes from a study in which participants were first asked to complete a questionnaire about various experiences in their lives. They had to say whether they had ever ridden a unicycle, visited a foreign city or wet the bed. After the participants had finished entering their data, they were told that a computer was busy preparing an automated biography of them – which the researcher then printed and handed to them.

In reality, the text was pre-planned in a way that would produce acute feelings of embarrassment. 'Although this student is not without faults,' it said, 'occasionally having some difficulties with bed wetting, he [or she] has continued to excel as a student at Cornell, and considers himself [or herself] to be a friendly, outgoing and caring person.' The texts also listed some hobbies and interests. As the participants were handed the document, they were told that another copy was going to be given to another student to evaluate. The participants then had to estimate how positively the new acquaintance would view them on a scale of 1 (much more negative than the impression of the average student) to 100 (much more positive than the impression of the average student). And the researchers really did give the introductory information to another student, and asked them to rate their impressions of how much they liked the person in question, using the same scale.

We can imagine the alleged bed-wetters' blushes as they read the printout, but the embarrassing information was interpreted far more positively than they predicted. The difference was particularly stark when the new acquaintances were equipped with additional statements about the student's hobbies and interests, in addition to the hints of nocturnal incontinence. With more details to process, they seem to have given surprisingly little weight to the slightly off-putting material; on the 100-point scale, they rated them at 69, an overwhelmingly positive evaluation. Remember that on this particular scale, the average student should receive a score of 50 – so this suggests that they were still highly inclined to make that person's acquaintance.[17]

Our egocentric thinking may lie behind these false expectations; we focus on the detail that is most salient to us – the source of our shame or embarrassment – while others are seeing the big picture. People are much less likely to think about your bed-wetting if they know that you are someone who regularly helps their friends, who aces their exams, or who has excellent musical taste. But we easily forget this, and expect that one embarrassing detail will override all other information.

In many cases, people will see a confession of vulnerability as a sign of authenticity. So even if the information that we provide is not, in itself, positive, it does at least show us that the other person is striving to create the joint understanding that is essential for a close bond to form, which could fuel people's desire to connect.

Dena Gromet and Emily Pronin at Princeton University asked students to imagine picking a few statements that might represent their inner life to a stranger. Some were asked to select from a list of fears and insecurities:

- I get frustrated easily and tend to give up on things before I should
- I'm overly critical of myself and often feel inadequate around others
- I can be closed-minded to ideas and opinions that are unlike my own
- I can be extremely impulsive and often regret the decisions I make

Others were asked to pick a few apt statements from a list of strengths, such as:

- I am pretty secure in who I am
- I am open to new ideas and opinions that are unlike my own
- I am level-headed and good at keeping my cool when making tough decisions

- I don't give up, and always try to see things through to the end

In each case, they were told that these statements would then be shown to another student, and were asked to answer the following question: How much on a scale of 1 (not very much) to 7 (very much) do you think this student would like you?

You can try it for yourself. What kinds of statements are more likely to foster connection? If you are like me, you would expect people to value the strengths over the weaknesses – and that's what the Princeton students thought, too. They assumed that their admissions of bad temper, closed-minded thinking and impulsivity would be rather off-putting.

To test whether those assumptions were justified, Gromet and Pronin next passed each set of statements on to a separate group of participants – and asked them to rate, on the same scale, how much they thought they would like the person who had written them. And their answers were the exact inverse of what the first set of participants had predicted. When reading someone's strength statements, this set of participants gave an average likeability rating of 3.8; when reading the vulnerability statements, they rated them as 4.3. As hypothesised, these differences were directly correlated to the participants' impressions of authenticity. When someone read the vulnerability statements, they assumed that the person who had written them was more genuine and honest, which then increased that person's likeability.[18]

Multiple psychological experiments have now replicated these findings. Whether it's declaring a secret crush, revealing our physical insecurities or admitting to a serious mistake at work, we assume that we will be judged harshly for our confessions. But many people will be more capable of empathising with our situation and appreciating our bravery in opening up.[19] Embracing our vulnerability can even benefit people in positions of power, who may strive to present a strong and flawless image to their followers. Leaders who disclose a potentially embarrassing weakness – such as shyness, anxiety about public speaking, or a fear of flying – score more highly on ratings of authenticity and engender greater loyalty from their workforce.[20]

Such findings chime with the qualitative research of the writer Brené Brown, who has spent years interviewing people about their feelings of shame. 'We love seeing raw truth and openness in other people, but we are afraid to let them see it in us,' she wrote in *Daring Greatly*. 'I'm drawn to your vulnerability but repelled by my own.'[21] We should all feel less shame for our imperfections, and the latest scientific findings should give us the confidence to bond over the beautiful mess of our lives.

EVADING, DODGING AND PALTERING

When we have a fact that we'd rather hide, we might hope that we can avoid self-disclosure without overtly lying. We may simply refuse to answer an awkward question, we can try to distract attention by focusing on a different point entirely, or we can make statements that are technically true but fail to address the relevant point. Unfortunately, each strategy comes with many risks, and few benefits.

Let's begin with evasion. Researchers at Harvard Business School asked participants to consider people's responses to (fictitious) dating profiles that explored some of the darker moments in the person's life – such as concealing a sexually transmitted disease from another person, stealing cash, making a false insurance claim, or having a fantasy of torturing someone. The profiles did not give in-depth details, but each member had (ostensibly) answered some multiple-choice questions about how often they had engaged in the relevant activity, with the following options: 'never', 'once', 'sometimes', 'frequently' or 'choose not to answer'.

Since these are purely unethical acts, I think it's safe to assume that most of us would prefer to meet someone with a perfectly clean record. But what if you had to choose between someone who admitted to a couple of these behaviours, and someone who opted not to answer? Who would you rather meet?

In the Harvard study, honesty almost always won out over evasion. People were considerably more likely to prefer someone who openly admitted to a sin to those who skirted the question. People who *refused to answer* whether they had ever hidden an STD from a partner, for instance, were judged just as harshly as someone who admitted to engaging in this behaviour 'frequently' and considerably worse than people who admitted to doing it 'sometimes' or 'once'.

If humans were strictly logical decision-making machines, this wouldn't make much sense. The person who had declined to answer may simply be a one-off offender, after all, so they'd be a safer bet than the person who had admitted to 'frequently' acting unethically. But we aren't calculators weighing up probabilities; we're interested in forming meaningful connections with people. Refusing to engage with the questions seemed to raise serious doubts about the person's overall integrity. Intriguingly, the team found much the same result when looking at the workplace: potential employers were more interested in hiring someone who had admitted to taking drugs, for example, than someone who chose not to disclose the information.[22]

Other strategies to evade the truth fare little better. We might choose to dodge the question by turning the conversation towards a less objectionable subject, for example – creating the illusion of answering when we've avoided the central point. This is a common tactic for politicians, and the research suggests that it can sometimes work if your answer is close enough to the original topic that inattentive listeners fail to spot your conversational side-step. But attentive listeners are still likely to notice the move, and they will judge you harshly for it.[23]

The related strategy of paltering – the use of technically true statements to create a false impression – is similarly risky. This technique is perhaps best illustrated by President Bill Clinton's words at the height of the Monica Lewinsky scandal. During an interview on PBS in 1998, the journalist Jim Lehrer asked a very blunt yes-or-no question: 'You had no sexual relationship with this young woman?' To most observers, Lehrer was asking whether Clinton had *ever* been involved with Lewinsky, but Clinton's response was carefully worded in the present tense. 'There is not a sexual relationship – that is accurate.' The answer may have satisfied some viewers, but although Clinton's statement was technically true – the affair was over by this point – he was clearly misleading people into thinking that the relationship had never occurred.[24]

Clinton's case is a particularly egregious example of paltering, but you have probably used the technique yourself in less serious settings. In a business negotiation, for example, you might be asked about forecasts for future growth, when you know that your sales have started to plateau. An honest speaker would be upfront about this fact, whereas a palterer might evade disclosure of current sales by focusing on *past* successes – 'Well, as you know, over the last ten years our sales have grown consistently' – without declaring that the growth is now over.

When asked about the best conversational strategy, most people prefer paltering to outright lying, and consider it to be more ethical. But it comes with severe repercussions when the other party realises that they have been deluded. Experiments show that people are just as negative about paltering as they are about any other kind of deception, and they will be less likely to trust that person a second time.[25] There is no good way to conceal facts without risking damage to our reputation, once the truth comes out.

THE MYTH OF THE WHITE LIE

Even 'prosocial dishonesty' – those untruths designed to save other people's feelings – can come with a penalty.

The concept of the white lie – so-called to emphasise the purity of the speaker's intentions – dates back to at least the sixteenth century, with a brief mention in the letters of Henry VIII's Lord Chancellor Thomas More. Writing to the Dutch theologian Desiderius Erasmus, More admits that he is not so 'superstitiously veracious as to reckon every white lie as black as murder'.[26] Like More, most people today consider white lies to be an essential social lubricant that helps conversation to flow more smoothly. Without them, you might assume that our interactions would simply grind to a halt or burn out with excess friction. But are these assumptions valid?

Emma Levine at the University of Chicago and Taya Cohen at Carnegie Mellon University concocted a series of studies to find out. In their first experiment, they recruited around 150 participants, who were divided into three groups. The first set was asked to be 'absolutely honest' in every conversation, at home and at work, for the next three days. 'Really try to be completely candid and open when you are sharing your thoughts, feelings, and opinions with others . . . Even though this may be difficult, you should do your absolute best to be honest.' The second set was told to be kind, caring and considerate for the same period, while the third was encouraged to behave normally. They then completed questionnaires about their experiences on each of the three nights of the experiment, and two weeks later they were asked to reflect on what they had learned from the experience.

When asked to predict who would fare better, most people assume that the participants who prioritised kindness would have the best experience, while the honest group would struggle to keep their friendships. Contrary to those expectations, however, the honest participants scored just as highly on measures of pleasure and social connection throughout the three days as those who were told to be kind. And they often found a lot of meaning in the exchanges. The negative information seemed to have been accompanied by deeper conversations, for instance – without damaging their overall relationship.

The participants' written accounts of the experiences, collected at the end of the study, speak for themselves. 'Being able to ask whatever question I desired and to answer in a truthful manner allowed me and the other person I was speaking [with] to be more open and comfortable,' one participant reported. They recalled an awkward interaction with one acquaintance, but felt that the authenticity of the exchange was worth the discomfort. 'What's the point of pretending when someone asks how you are feeling?' this person added. 'I feel that, generally, being honest allows for better relationships and more trust. Since that experience, I have been trying to be more honest in my daily life.' Others reported a sense of liberation. 'People reacted differently than what I thought. They liked and appreciated the honesty and I did not believe that would happen. It was refreshing, meaning that I was happy to talk about what was on my mind and not worry about what was said.'

In a follow-up experiment, Levine and Cohen asked pairs of friends, colleagues, roommates or romantic partners to open up about personal issues – such as the last time they cried or their most embarrassing moment – with as much honesty as possible. In each case, the honest communication triggered meaningful conversations that were not nearly as difficult as people had feared, and the benefits of the candid disclosure on their overall wellbeing continued for at least a week after the intervention had ended. Finally, the researchers also asked people to share negative feedback with someone close – 'one thing you think this person should do differently, change about themselves, or improve upon'. Once again, the reactions were much better than expected.[27]

This seems to be a further example of adopting an overly narrow focus in our view of the situation.[28] We concentrate on the negative content of the feedback and its potential to hurt while forgetting that our friend will consider our comments within the overall context of the relationship. If we have already shown ourselves to be kind, our friend will have many memories of our past behaviour, which should reassure them that we have their best interests at heart. That knowledge may be enough to soften the blow of the information we are delivering; at the very least, it will prevent them from 'shooting the messenger'.

When Levine questioned people more generally about their perceptions of 'prosocial deception', she found that white lies are tolerated under very limited circumstances: namely, when the situation is beyond someone's control and declaring the truth could bring no benefits. To give an obvious example: there's no need to tell a bride or groom that they look terrible on their wedding day, if they can do nothing to change their appearance before they walk down the aisle. When the truth will help to save future embarrassment, however, honesty is the best policy, while white lies are considered paternalistic and patronising.[29]

On a practical level, kind fibs can stunt learning and growth. Researchers at Harvard Business School, for instance, asked pairs of participants to take part in a public speaking competition, with cash prizes. One was the speaker and one acted as a coach, giving feedback on their rehearsal. Most of the coaches underestimated how much their partners would want honest feedback, and they tended to give more positive encouragement than suggestions for improvement. And that came at the cost of performance; the less constructive criticism that someone received during the practice session, the worse the judges rated their final speech. The incentives here were relatively small – just $50 for the winning speaker and $25 for the winning coach – but an aversion to honesty could result in a failed sales presentation or a poor performance at an interview that loses someone their dream job. Troublingly, further studies have shown that the more serious the potential consequences, the less willing people are to provide truthful criticism.[30]

These difficult conversations need to be handled with tact. Clearly stating your positive intentions before delivering a potentially painful truth can help to soften the blow of bad news or criticism, and you can make your best effort to phrase it in the most constructive terms possible, so that it's clear how the person will be able to learn from what you say. To tell someone that their presentation is 'boring', for example, is no more useful than telling them – falsely – that their presentation is 'brilliant'; it's simply more hurtful. Providing specific examples of the places where you lost attention, and articulating the specific reasons that you didn't find it relatable, would provide much more useful pointers to improvement. In everyday language, we often confuse being honest with being 'blunt' – but when you are delivering truthful feedback, it's more important than ever to talk with precision and nuance. Finally, you might offer to help the person deal with the fallout – either emotionally or practically. (In this example, you could offer to go out for coffee to offer more general writing advice.)

One well-meaning – but largely futile – strategy is the 'feedback sandwich', in which the bitter pill of criticism is hidden between fake compliments. (For this reason, you might have also heard it described as the 'shit sandwich'.) This devalues the praise and doesn't make the negative feedback any more palatable – a lose-lose situation.[31] It's better to present the truth frankly than to wrap disappointing news in hollow words of comfort.

TRUST, AND BE TRUSTED

We have now seen abundant evidence for the seventh law of connection. As you attempt to be a little more honest and open about your own life, you might also consider whether you offer a safe space for others to do the same. Your love of gossip, for instance, might often make for entertaining conversations, but if you engage in it too often it will lead others to think twice before sharing their secrets. They'll filter their life for you, resulting in missed opportunities for bonding.

When others do open their hearts, you can validate their feelings by showing active care and interest – but do not assume that the topic is fair game for any future conversation. Someone may find relief in disclosing a secret, but that does not mean that they will want to discuss it at every opportunity. Indeed, the possibility of being reminded of our vulnerabilities is often a key consideration when deciding whether to reveal personal information.[32] So tread carefully before you choose to broach the subject again. In many cases, it may be best to let your acquaintance know that you are willing to discuss their situation whenever they feel comfortable, and to then let them take the lead.

For similar reasons, you should check your responses when receiving bad news or negative feedback. A quick flash of anger at the first sign of criticism will only discourage others from sharing honest insights in the future. If they are genuinely offering a constructive opinion, try to remind yourself of their good intentions and turn the conversation to the ways that you might be able to use the viewpoint they have provided.

Finally, you might consider your perceptions of other people's trustworthiness, since this will powerfully influence your responses to others' self-disclosure and your willingness to react in good faith to the words they say. How would you rate, on a scale of 1 (completely disagree) to 5 (completely agree), the following statements:

- Most people are basically honest
- Most people trust a person if the person trusts them
- Most people are basically good-natured and kind

This is known as the 'generalised trust scale'. If you are of a cynical disposition, you might expect higher scorers to be gullible and easily manipulated – but you'd be wrong. People who are trusting seem to be the much better judges of character.

In one striking study, Canadian researchers asked a group of MBA students to go through mock interviews for a real job opportunity. Half were asked to tell nothing but the truth, while the others were given a green light to lie about their credentials as much as they liked. The interviews were filmed and then shown to a second group of participants, who offered their perceptions of the students' honesty, along with their opinions of whether they should be hired for the job. It turned out that the more trusting participants were more accurate than the cynics, who simply weren't as good at picking up on the 'tells' that can signal when someone is being deceptive, and were more likely to endorse picking deceitful candidates for the job.

It seems that, with their more unguarded attitudes, the more trusting participants had learned to be more sensitive to cues, such as changes in someone's voice, that might signal truth or deception. Throughout their lives, these people may have actively put themselves in positions where they had to rely on the honesty of others – training them to be better judges of character. The cynics, in contrast, may have been so worried about being fooled in the past that they had limited their interactions with others. As a result, they simply hadn't been paying enough attention to human behaviour to differentiate between people acting in good or bad faith. Higher levels of trust may increase our social intelligence, so that we are better able to navigate the complex behaviours of the people around us.[33]

Being open and honest with others – and trusting them to act the same in return – is not always going to be easy. Like Schopenhauer's porcupines trying to huddle in the cold, we face the danger of injury every time we move closer to another person. And yet the benefits far outweigh those risks. When you learn to act with good faith, you will find that your candour and trust are often rewarded with appreciation and connection, and the world is a little less hostile than it once seemed. The spines that hold us apart are often illusory.

What you need to know

- Secrets come with a sense of physical and mental burden, and social isolation – all of which impact our wellbeing
- When we feel confident to do so, spontaneously revealing a secret to another person can enhance the sense of connection for both parties
- People tend to view our insecurities and embarrassments far more positively than we believe – a phenomenon known as the beautiful mess effect
- Attempts to evade the truth by dodging questions or paltering tend to be viewed as badly as outright lying
- White lies, told to save hurt feelings, can appear paternalistic and patronising. They are only acceptable if the person has no control over the situation in question

- We consistently underestimate people's desire for honest feedback – and this bias is particularly pronounced in the situations where truthful insights could have the largest consequences

Action points

- Spend a week attempting to address every social interaction with total honesty. Keep a diary, and note whether your conversations and email exchanges tend to be more liberating and enriching than you had originally expected
- When offering feedback, avoid the 'shit sandwich' of positive encouragement enclosing the negative information. It devalues the praise and does little to sweeten the filling
- Avoid blunt statements and instead focus on the precise details of the problem. Where possible, suggest positive steps that might resolve the issue
- Show others that their secrets are safe with you by avoiding malicious gossip. After someone has self-disclosed, be careful not to broach the painful topic too often – but make it clear that you are ready to talk again whenever they feel comfortable

CHAPTER 8

AVOIDING ENVY AND ENJOYING CONFELICITY

If you have spent any length of time on social media, you'll be well acquainted with the 'humblebrag' – a seemingly self-deprecating comment or complaint that simply draws attention to someone's beauty, wealth, education or professional success: *I can't even count the number of people who told me I look like a celebrity. Like really* or *Graduating from two universities means you get double the calls asking for money/donations. So pushy and annoying!*

Humblebrags are particularly popular with celebrities. Take the actor Jared Leto's tweet after being recognised by a top fashion magazine for his sharp dress sense:

> Just won GQ style award in Germany. Obviously they made a mistake. I wonder how long till they come take it back. ;) #andthewinnerisWHOOPS![1]

Or Meryl Streep's speech while she was accepting her third Oscar:

> When they called my name, I had this feeling I could hear half of America going, 'Oh, no. Oh, come on, why? Her. Again.' . . . But. Whatever.[2]

The humblebragging habit is not limited to the English-speaking world. On Chinese social media, users denigrate certain posts as 'Versailles literature'.[3] The term comes from the manga series *The Rose of Versailles*, which depicts the lavish lifestyle of Marie Antoinette, and it is now used to describe any posts that show off someone's riches and social status while feigning nonchalance or disappointment, such as:

> There weren't enough electric car charging stations in the neighbourhood and we weren't allowed to install new ones. So we had no choice but to move to a bigger house with a private garage for my husband's Tesla.[4]

We may smirk, but the motives for humblebragging are no laughing matter. They are a misguided attempt to overcome one of the major dilemmas in our social interactions.

It is natural to want respect from others: in human evolution, our reputation among our peers would have been essential for our survival. People who were known to be smart, hard-working, generous and cooperative would have been valued group members and may have found it easier to form useful alliances with their peers; they would have also been more attractive mates. One person's rise can lead to another's downfall, however, creating resentment among people who are struggling to gain the same kind of status. In evolutionary history, our ancestors would have been particularly suspicious of people who seemed to be getting too cocksure without paying their dues – those dubious individuals who become too grand to pull their weight or contribute to the greater good. As a result, we evolved emotions such as envy, which incites us to punish those who might be garnering too much attention without merit, while inspiring us to strive for the same achievements ourselves.

Whenever we display joy about a new relationship or talk about a promotion at work, we risk triggering these unpleasant feelings in the people we want to impress. That is a major barrier to social connection and could help to explain why so many successful people also feel incredibly lonely. The humblebrag attempts to offset the resentment with humour or sympathy, but recent research suggests it may only amplify the emotions it is designed to avoid. Fortunately, there are many other ways to be open about our successes. In place of resentment and malicious envy, they evoke 'confelicity' – a shared joy in each other's achievements that strengthens our connections. For our eighth law of connection, we will learn why we should openly celebrate our achievements without fearing envy.

BRAGGING RIGHTS

Before we examine the psychology of false modesty and the reasons it backfires so badly, let's consider the perils of self-praise more generally. You do not need me to tell you that showy displays of unfounded arrogance can lead to embarassment. As William Shakespeare wrote in *All's Well That Ends Well*, 'it will come to pass/ that every braggart shall be found an ass'.

Psychological research suggests that our brags are judged on two key dimensions. The first is accuracy. Students singing their own praises will be judged far more positively if they can mention some proof of their academic potential – such as a high grade in a recent exam – than if they offer no facts to back up their claims, for instance. And they will be judged especially harshly if there is evidence to the contrary: if they have not only failed to excel but performed poorly.[5] This makes sense from the evolutionary standpoint: if there's one thing worse than someone attempting to climb the social order and vaunt their newfound status, it is doing so dishonestly, since that instantly raises questions about their overall conduct. It suggests that they may resort to any means to gain approval.

Many of us understand this intuitively, but some will still be tempted to brag about non-existent successes when they think that the truth cannot be found out, as Barry Schlenker at the University of Florida showed in a classic study from the 1970s. The participants were first given a series of individual aptitude tests, before taking part in a group exercise. This took the form of a quiz, where each member could vote for the correct answer. In some teams, the votes were counted anonymously in a secret ballot; for others, they would have to reveal their answers to the whole group, even if they were wrong.

Schlenker was interested to see whether the prospect of anonymity would change the way that the participants introduced themselves to the other team members – and he was not disappointed. When they were allowed to cast their votes in the secret ballot, the participants were far more likely to exaggerate their intellectual prowess, compared to those who had to make their answers public. They knew that their claims could never be checked against their actual performance, and so they were far happier to brag. This was even true of the people who had been told that they had performed very poorly in the first session's aptitude tests: when there was no chance of being proved wrong, the participants were happy to boast about a talent that didn't exist.[6]

In real life, it can be very hard to tell when our true abilities and achievements might eventually come to light. We may brag to our families about our popularity with our colleagues, thinking that the two social spheres will never come into contact – only to discover that our cousin happens to go to the same gym as our co-worker. Or we may tell our colleagues that we are fluent in a language, only to find ourselves red-faced when we are asked to interpret a native speaker.

Worse still, our judgements of our abilities are often deeply flawed, meaning that we may not even realise how misleading our boasts have been.[7] And it is often the people who are least talented who are the most over-confident – a tendency that is named the Dunning–Kruger Effect after the two researchers, David Dunning and Justin Kruger, who discovered the phenomenon.[8]

Journalists and comedians have been quick to apply the term to many a politician's ludicrous boast, but the truth is that most of us share this tendency to overestimate our skills. We must be confident that we have sufficient bragging rights, otherwise we all face the risk of being 'found an ass'.

THE HUBRIS HYPOTHESIS

In addition to their accuracy, our boasts will be judged on whether or not we are making a direct comparison to other people. According to a theory known as the 'hubris hypothesis', people are far more likely to object to self-praise if it implies a negative judgement of others, and if we avoid such comparisons, our boasts will be viewed far more favourably.

To get a flavour of this research, consider the following statements:

1. 'You know, I am a better person to be friends with than others . . . I am more often ready to have a ball . . . I also do more for people who belong to my circle of friends than others do . . . If I compare myself to others, I may well say that I'm more devoted, loyal and open-minded and that you can have more fun with me.'

2. 'You know, I am a good person to be friends with . . . I am often ready to have a ball . . . I also do a lot for people who belong to my circle of friends. If I look at myself, I may well say that I'm devoted, loyal and open-minded and that you can have a lot of fun with me.'

Which of the two people would you prefer to meet?

With all the goodwill in the world, you could not describe either of the above statements as modest: they are celebrating desirable traits and behaviours that others would love to demonstrate. The only difference is whether the person presents those alleged assets relative to other people's traits – as in the first example – or whether they avoid explicit comparison, as we see in the second example. Across multiple studies, Vera Hoorens at the Katholieke Universiteit Leuven has shown that people making direct comparisons ('explicit' braggarts) are much less likeable than those who simply big themselves up without any reference to others ('implicit' braggarts).[9]

Why on earth should this be the case? When considered logically, it makes no sense, but Hoorens's further research suggests that the answer revolves around our fear of judgement. If someone explicitly makes social comparisons about others *in general*, we assume that they will have a negative view of *us personally*. This leads us to feel our own status is threatened, and so we instinctively recoil from them.[10] Our aversion to direct social comparison can be so strong that it leads to hostility and aggression. When considering different university dormmates, for instance, participants wanted to assign the explicit braggarts the least appealing household chores, and they were happy to sabotage the explicit braggarts' chances of winning a competition for a monetary prize.[11]

You might expect that claims of equality would be preferable to open brags. If modesty were always considered a virtue, someone who writes 'I am *as good a* person to be friends with as others are . . . I am *as often* ready to have a ball' should be treated more warmly than the implicit braggart who claims 'I am a good person to be friends with'. This doesn't appear to be the case, though. While people who claim equality are consistently considered more likeable than someone who places themselves above others, they are rated no more favourably than people who make implicit brags. There appears to be no harm in blowing your own trumpet, as long as you don't make the explicit claim of superiority that triggers people's fears of being judged negatively.[12]

Such findings should be relevant to any conversation. Whether you are showing off your new house, celebrating your skills as a parent or describing the contents of your workouts, there's no need to bring other people into it. Understanding this should be of particular use in the workplace. Self-presentation can make a big difference to your prospects of promotion, and as you attempt to prove yourself, it may feel tempting to describe how you rank compared to your colleagues, but such overt comparisons will backfire. Be content to describe the details of your individual accomplishments and let your boss do the maths.

Whether you regularly fall for this trap may depend on your

personality.[13] The trait of narcissism is measured by asking people to consider pairs of statements and then select the most applicable, such as:

- *I like to be the centre of attention* or *I prefer to blend in with the crowd*
- *I insist upon getting the respect that is due me* or *I usually get the respect that I deserve*
- *I am more capable than other people* or *There is a lot that I can learn from other people*

These items come from the Narcissistic Personality Inventory, and if you found yourself agreeing with the first statements in each pair, you may have higher-than-average narcissism.[14] Don't be too disheartened: narcissists are perfectly capable of developing close relationships, but you may benefit from a little more awareness of your need for validation and respect. Due to their need for respect, narcissists are much more prone to talk themselves up while putting others down, and they are less conscious of the ways that their words could make others feel. As a result, they may be especially likely to make the social comparisons that can be so alienating.

If you identify as a narcissist, you may need to make a bigger effort to avoid this pernicious habit. You may want to be the tallest poppy in the field, but excessive social comparison will prevent you from making the impression you want.

THE PERILS OF FALSE MODESTY

This research should be reassuring; by following these simple rules, we can celebrate our successes while avoiding much of the social censure that normally comes from unfounded bragging. Yet we may still wonder whether it would be better to simply downplay our achievements. Surely people will respect our humility over someone who acts as their own hype man? This is risky, however, since false modesty is simply another form of the 'pro-social lie' – and as we have already seen, lies like this can harm people's perceptions of your authenticity. As the seventeenth-century French philosopher Jean de la Bruyère famously noted, the desire to appear modest is itself a sign of extreme vanity, and people are unlikely to warm to the attempted deceit.[15]

This may explain why humblebrags are so off-putting. Analyses of people's Facebook posts show that humblebragging reduces perceptions of sincerity, compared to straightforward brags.[16] Humblebraggers may believe that the self-deprecating comments accompanying their boasts make them look relatable or down to earth, but it's more often seen as a sign of manipulation.

Attempts to hide our success from others can backfire for similar reasons, as the behavioural scientist Annabelle Roberts has shown in a series of recent studies. In a preliminary survey, she found that more than 80 per cent of people admit to concealing good news – such as impressive exam results, a promotion at work or flashy consumer purchases – from the people around them. When asked to elaborate, most of these people reported good intentions: they wanted to avoid provoking jealousy or creating awkwardness in a conversation. Yet Roberts's subsequent experiments demonstrated that keeping these things secret rarely had the intended effects. Rather than communicating humility, hiding an achievement or stroke of luck betrays low expectations of the other person's self-worth, and suggests that you may be highly conscious of a difference in status between you.

In one study, Roberts's team recruited around 150 pairs of participants who were already known to each other – as friends, romantic partners, family members or colleagues. After providing basic information about their relationship, they asked one member of each pair to describe a recent success that aroused pride, but which they hadn't, for whatever reason, yet told their partner in the experiment. Their responses ranged from cooking an ambitious meal or reaching a personal best on their weightlifting to getting a perfect grade at university or attracting a large number of social media followers – the kinds of achievements that might put us all in a great mood for a few days.

Having collected these statements, the scientists next turned their attention to the friend, colleague or romantic partner (labelled the 'target' in the write-up of the experiment). The researchers revealed the question posed to the first member of the pair, before providing false information about whether or not they had chosen to keep their response a secret. If the target was told that their associate had chosen to hide the response, then it remained unknown to the target; if they were told that their associate had chosen to share their success, then the target read the text. In each case, the target was given a series of questions probing how they now felt about their associate. They were also given the option of spending up to $1 on an e-card as a token of appreciation for the person in question, or of keeping the money to themselves.

You might expect that feelings of jealousy arising from the self-promoting statement would cast a shadow over the relationship and reduce the participants' sense of connection – yet hiding success evoked the worst responses on almost every measure. The targets reported feeling distinctly insulted by the decision and they felt less close to their associate as a result. This, in turn, shaped their behaviour. When given the chance to send an uplifting e-card to their associate, they tended to shun the opportunity and instead collected the dollar themselves: a small but petty gesture indicating that they were simply less inclined to celebrate their relationship.

This was, admittedly, a rather artificial set-up, but subsequent studies asked participants to consider various scenarios from day-to-day life and to report how they would react in similar situations. In every case, people who hid their successes were judged harshly. Imagine, for instance, that your mother tells you that your brother has recently been promoted and earned a pay rise of £20,000. How would you feel if you then bumped into your brother and he failed to mention his new circumstances, as if his life had not changed since the last time you met? If you are like most people in Roberts's study, you would feel insulted and alienated by your brother's avoidance of the topic.

The sour feelings seem to emerge from the motives that we ascribe to the person hiding their achievements: we assume that they are acting paternalistically. Roberts, for instance, asked participants to consider the story of two co-workers who are both looking for a new job. One colleague, named Alex, receives the chance to give a presentation to a potential employer, but neglects to tell his friend during their next meeting, even though they discuss their job hunts. There could be multiple explanations for this behaviour – including sheer forgetfulness – but the participants saw Alex's behaviour as patronisingly protective. Alex, they assumed, would be hiding the truth because he thought they 'couldn't handle the truth' and would feel threatened by his success. They sensed he was trying to manage their emotions, which feels manipulative.

The researchers were interested to see if this interpretation would erode people's overall trust in Alex. To do so, they asked the participants to rate statements such as 'Soon after this interaction, I would feel comfortable sharing my most outlandish ideas and hopes with this person' and 'Soon after this interaction, I would feel comfortable admitting my worst mistakes to this person'. Their hypothesis was spot on. When Alex had hidden this potential job opening from them, the participants predicted that they would be far less likely to engage in self-disclosure themselves, and they were less willing to collaborate with him in the future.

As you would expect, people's reactions depend on the specific path the conversation takes. In the above example, we might ask Alex about his career prospects directly ('How are your job applications going?') or indirectly ('What's new with you?'). Neglecting to mention the good news is hurtful in each case, but the sense of alienation would be particularly acute if we had provided an obvious opportunity for him to speak about his good fortune. In such circumstances, it is almost impossible not to feel shut out from someone else's life.

Few people want to be treated like a fragile child who is set to have a breakdown at the slightest threat to their ego, but we seem to forget these risks when we choose to hide our successes from others. Roberts's team also examined the effects of hiding success on social networks and whether the closeness of the relationship matters. Across all these experiments, the conclusions are the same: you would do better to be open about your achievements than to hide them.[17] Provided that we choose our words tactfully and avoid comparing ourselves directly to others, people's feelings are not nearly as delicate as we fear.

MITFREUDE, OR CONFELICITY

To do justice to these findings, we may need to update our understanding of empathy. When scientists investigate shared emotions, their experiments have traditionally focused on negative feelings: our capacity to feel another's pain or distress. The word compassion literally derives from the Latin for 'shared suffering', after all, and in a world filled with cruelty, it might have seemed sensible to explore the ways in which we can increase concern for others. Confelicity – shared joy at another's success or wellbeing – has received much less attention, but it may be no less important to our relationships, as an essential means of providing and receiving social support.

Friedrich Nietzsche thought as much. You may be surprised to hear of this sentiment coming from the German philosopher – in the popular imagination, Nietzsche often features as a tormented genius and a staunch individualist. Yet he also coined the term *Mitfreude* – 'joying with' – that provides a German cousin to English's confelicity, and the saintly sister of *Schadenfreude*, our joy (*Freude*) at another's misfortune (*Schade*).

'The serpent that stings us means to hurt us and rejoices as it does so; the lowest animal can imagine the pain of others,' he wrote in *Human, All Too Human*. 'But to imagine the joy of others and to rejoice at it is the highest privilege of the highest animals, and among them it is accessible only to the choicest exemplars – thus a rare humanum: so that there have been philosophers who have denied the existence of *Mitfreude*.' As a foundation for friendship, Nietzsche considered *Mitfreude* to be even more important than compassion. '*Mitfreude* increases the force of the world,' he wrote elsewhere. 'Fellow rejoicing, not fellow suffering, makes the friend.'[18]

Whatever name you give the phenomenon, recent psychological research suggests that confelicity is more common than we might have estimated, and hostile reactions are far less likely. And if you already have mutual trust and respect, the emotional contagion of confelicity provides ample opportunity to build on those feelings.

Multiple experiments show that when someone else responds to our good news with feelings of excitement and enthusiasm, it amplifies our good mood, and through the creation of shared reality their validation will make the event more meaningful and memorable. The person hearing the news, in contrast, feels valued by the fact that we wish to share the experience with them. By openly revelling in the things that have made us proud, joyful or excited, we have demonstrated our trust that they have our best interests at heart. Like other forms of self-disclosure, this reveals our authenticity and intention to build an even stronger bond; it is an invitation for them to take part in our inner life – and that is intensely rewarding to experience.[19]

The result of confelicity is increased closeness on both sides – as measured by the Inclusion-of-Other-in-Self scale on p. 37. The more that romantic couples rejoice in each other's moments of pleasure, for instance, the greater the overlap of the concentric circles representing their individual identities, which then translates into higher ratings of relationship quality.[20] As you would expect from the enhanced sense of connection and acceptance, regular doses of confelicity brings better emotional wellbeing. When we avoid sharing some good news with the people we love – through a lack of confidence that they will care or the fear of seeming boastful – we deprive ourselves of all those benefits.[21]

Many of us currently miss the opportunity for confelicity. Researchers at the University of Michigan asked participants to consider the following scenario: 'Imagine that you have recently received a promotion at work. Around this time, you go out to dinner with a close friend. During your dinner conversation, your friend asks you, "How is work going?"' The participants had to give their answer to the question before predicting their friends' reactions on various psychological scales. You might think that the good news would be most obvious topic of discussion, but 40 per cent of the participants declared that they would not tell their friend about the promotion, due to fears of appearing like a braggart. But the research shows that sharing their excitement and happiness would only bring them closer to their friend.[22]

Even if our declarations do incite some jealousy, it does not always come with a bitter flavour: it can be seen as a motivation to change our circumstances and strive for the same success. Psychologists describe this as 'benign envy', as a direct contrast to the 'malicious envy' that might lead to rancour and resentment. In evolution, benign envy would have helped us to protect our position in the group hierarchy without leading to infighting or recriminations. I know from my own experience that seeing a friend succeed, where I have failed, can bring some pangs of regret that I was not standing in their place, but that does not have to override my genuine pleasure at seeing their happiness. Humans are complex creatures who can experience multiple emotions at once – and vicarious joy will often win out.

THE JOURNEY

Drawing on all these findings, our eighth law of connection is as follows: **Do not fear envy. Disclose your successes but be accurate in your statements and avoid comparing yourself to others. Enjoy 'confelicity'.** It has to be practised with care and sensitivity. You don't need to declare your engagement at the funeral of your friend's spouse, or show off your new car on the day that your sister loses her job. Even in the best circumstances, self-celebration is best enjoyed in moderation. There will often be plenty of other topics of conversation that are of much greater interest to your audience.

Whenever you feel tempted to engage in self-praise, you should ask yourself whether this fact is personally important to you, and whether it will help the other person to understand you better. Do the people around you really need to know that you've been mistaken for a model to enter a shared reality, for example? Or are there more relevant facts to share about yourself? If your only motive is to highlight your high social status, without providing any particular insight into your thoughts, feelings or dreams, you would do better to keep your mouth shut.

When a moment of personal happiness provides a genuine opportunity for self-disclosure and mutual understanding, however, this recent psychological research can offer some useful guidelines to make the most of the experience. Your primary considerations should be to demonstrate the authenticity of your feelings and your genuine desire to build social connection. In many cases, you should let people know as soon as possible to avoid any suspicions that you have been withholding the information in a misguided attempt to manage their feelings. While describing your success, you must be honest about the way that you feel, but do not exaggerate your joy or make any kinds of claims that are unfounded. And you must avoid overtly comparing yourself to anyone else. As we have seen, we may be forgiven for declaring our own genius, but we should do so in absolute rather than relative terms.

If you still fear that your achievements will invite malicious envy, you may find it helpful to discuss the journey that led to the success, including some of the failures along the way. If you have just published a bestselling novel, for example, you might take a leaf out of J.K. Rowling's book and describe all the rejections you faced before it finally found a publisher: Rowling has frequently described how more than a dozen publishers declined *Harry Potter and the Philosopher's Stone* before it became one of the most widely read books of all time, and she has even shared some of the rejection notes that she received.[23]

Revisiting periods of difficulty during a time of great success may seem dangerously close to a humblebrag. To avoid stepping over that line, you need to describe specific disappointments and setbacks. Acknowledging a series of bruising rejections – an experience that many struggling writers will have to face – is very different from moaning about the number of luxurious Tesla cars in your garage or your 'embarrassment' at being voted one of the world's sexiest men. And there are good reasons why it works. For one thing, your descriptions of the genuine challenges you faced will make it easier for other people to comprehend the reasons you are so proud of what you have achieved, contributing to the mutual understanding between you. It is also of practical benefit, offering genuinely useful information for people who may wish to follow in your footsteps. Your achievement is transformed from a source of jealousy into a wellspring of inspiration.

Such benefits were evident in a study of entrepreneurs vying for investment in a 'pitch competition'. When someone owned up to their mistakes or failings and described the lessons they had learned, the other competitors felt less malicious envy towards them, and greater benign envy. They responded more positively to statements such as 'I will try harder to obtain funding for my start-up at the next opportunity' and 'this entrepreneur's success encourages me', for instance – suggesting that they saw their peer's success as an inspiration rather than a threat.

Crucially, the acknowledgement of failure, alongside declarations of success, contributed to perceptions of *authentic* pride while reducing perceptions of hubris. The participants tended to rate these people as being 'fulfilled', 'accomplished', and 'confident', 'like he or she has self-worth', for example – just the kind of impressions we might hope to give at a moment of success. And the participants were significantly less likely to consider them 'conceited', 'snobbish', 'egotistical' or 'smug'. You might wonder whether the admission of failure could come at the cost of overall respect, reducing perceptions of competence or skill, but this was not the case. Participants' admiration for the individual's achievements was just as high when they knew about the bumps in the road to that goal as when they only heard about the person's successes.[24]

Such discoveries go against centuries of philosophical doctrine. 'If you want people to think well of you, do not speak well of yourself,' Blaise Pascal told us in his *Pensées*.[25] Millions of us have grown up with this belief, but it is wrong. Provided that we are open about the difficulties that we have faced, we should have no fear of expressing our achievements; confelicity is an essential ingredient of connection. Our happiness can be contagious.

What you need to know

- Self-praise is more warmly accepted if we can back up our claims with factual evidence. Bragging about your athletic ability will be better tolerated if you can also describe a recent triumph at half-marathon, for instance
- Making direct comparisons to others suggests that you are overly preoccupied with hierarchy. This triggers people's fears of being judged negatively, reducing your likeability and leading them to act with hostility towards you
- False modesty screams insincerity, which makes you less likeable than someone who boasts overtly. This is particularly true for humblebrags that try to make a complaint out of an achievement
- When you share your successes, people feel valued and trusted. By owning your pride and expressing it to others, you can create a shared experience of 'confelicity' that is a potent source of wellbeing

Action points

- Question the motives behind your self-praise. If you simply wish to underline your status, it may be better to keep quiet. If the information will genuinely help others to understand your experience and mental life, then feel free to share your story
- Choose your words carefully so that you cannot be seen to exaggerate your talents or good fortune, without downplaying the facts. Where appropriate, recognise others' contributions

- If you are discussing a recent success, mention some of the challenges that you faced to reach your goal, and acknowledge the mistakes that you made along the way. This reduces malicious envy
- Remain conscious of the conversational balance. People may run out of confelicity if you stop paying attention to their needs

CHAPTER 9

ASKING FOR HELP

Few people in history have demonstrated such refined social skills as Benjamin Franklin, whose capacity to build bridges between opposing parties would change the landscape of American politics. He negotiated the Treaty of Alliance in 1778, formalising French support for the newly established United States in the Revolutionary War, and – five years later – he set out the terms of US independence from the British Empire. As an elder statesman, Franklin would play a pivotal role in the drafting of his country's Constitution, settling long-standing disputes between conflicting factions that had threatened to tear the union apart.

Every illustrious career has to begin somewhere humble, however, and in 1736 Franklin was just embarking on his political journey as a lowly clerk in Pennsylvania's General Assembly. He was generally favoured by the other members and was initially selected for the role without opposition. The post was renewed annually, however, and the following year Franklin faced a rival who made a long speech against him. Franklin prevailed and was reselected, but he suspected that his rival's antipathy could thwart his future ambitions if it continued, and so he decided to win him over. Shunning the temptation of 'servile respect', he instead decided to ask the man a favour. He was inspired, he later wrote, by the old saying, 'He that has once done you a kindness will be more ready to do you another, than he whom you yourself have obliged.'

The two men's shared love of literature offered a perfect opportunity to put the tactic to the test. 'Having heard that he had in his library a certain very scarce and curious book, I wrote a note to him, expressing my desire of perusing that book, and requesting he would do me the favour of lending it to me for a few days,' Franklin explained in his autobiography. The man obliged, and Franklin returned it a week later with an enthusiastic thank-you note. 'When we next met in the House, he spoke to me (which he had never done before), and with great civility; and he ever after manifested a readiness to serve me on all occasions, so that we became great friends, and our friendship continued to his death,' he reported.[1]

We can all learn a lot from these observations. Requesting a favour is rarely easy, yet psychological research suggests that most people are considerably more generous than we expect, and in many instances our requests can even *increase* their esteem for us as Franklin observed. This will lead us to our ninth law of connection: **Ask for help when you need it, in the expectation that your pleas for support can build a stronger long-term bond.**

With any power comes responsibility, however, and if we want to form genuine and long-lasting relationships, we must be careful not to abuse our influence. Knowing how to phrase our pleas, so that the other person can make a genuine offer of assistance without feeling pressured or unduly inconvenienced, is one of the most important skills we can learn to enhance our social lives.

WHY DIDN'T YOU ASK?

Shyness can be a serious barrier to success. In education, students who are too shy to ask teachers for assistance in their studies may flail around in confusion and frustration, or they become disenchanted with learning. Either way, they are considerably more likely to fail their exams than students who will more confidently request support when it is required.[2] This continues into the workplace. While a certain level of autonomy is essential for most professions, people who struggle by themselves without asking for any assistance perform worse in the long term, and their wellbeing suffers from the lack of support.[3]

In extreme cases, our fear of troubling others can be a secret assassin. Consider the frustration of the US doctor Henry J. Heimlich, who invented the famous manoeuvre to remove objects lodged in the windpipe. He noted that 'sometimes, a victim of choking becomes embarrassed by his predicament and succeeds in getting up and leaving the eating area unnoticed. In a nearby room, he loses consciousness and, if unattended, he will die or suffer permanent brain damage.'[4]

You might hope that people in positions of high responsibility would have overcome this aversion to calling on others for support, but this cannot be taken for granted. In 1990, Avianca Flight 052 from Bogotá, Colombia, ran out of fuel and crashed into Long Island. Transcripts of the accident revealed serious miscommunication between the pilots and air traffic control at JFK airport, as the first officer failed to convey the need for an *immediate* landing. Not recognising the nature of the emergency, the controllers kept them waiting for a runway until eventually the plane spiralled into the hillside of Cove Neck, New York. Some of the first officer's final words to his captain were 'the guy is angry'; he feared he'd offended his colleagues by being too pushy. Post-mortem analyses of the crash concluded that the passengers and crew could have all been saved if the pilots had made a more assertive request for help.[5]

Such dramatic stories are rare, but in everyday life our aversion to help-seeking prevents our friends and family from relieving our stresses, which will have consequences for our health. If we want to overcome this barrier, we must first understand where our fear comes from and in what situations such anxieties may be merited.

Vanessa Bohns at Cornell University has led much of this research. She was inspired by personal experience. As a graduate student in New York City, she was asked to collect survey data in Penn Station for an academic research project. Each time she approached a passer-by, she expected to be greeted with exasperation. Yet the hostile responses rarely came; many more people were willing to answer the questionnaires than she had expected. Bohns began to wonder whether this was another cognitive bias – that we consistently underestimate others' willingness to cooperate – and over the following decade she conducted multiple studies that have confirmed that this is indeed the case.

One experiment directly replicated her own experience in Penn Station: the participants had to approach strangers on the university campus and ask them to complete a survey. To get five responses, most people estimated that they'd need to ask at least twenty people. In practice, that number was closer to ten.[6] In another experiment, participants leaving the lab had to ask for directions to Columbia's campus gym, the entrance of which is below street level and difficult to spot. If the passer-by gave directions, the participant then had to ask if their new acquaintance would walk them to the gym. 'I was just over there, and I couldn't find it. Can you take me there?' they explained. The participants asked for this favour around three blocks away from the gym – meaning that the stranger would need to make a small but inconvenient detour to fulfil the request. Suspecting that most people would be unwilling to take time out of their day to help another, the participants assumed that they'd have to approach about seven people before someone would agree to take the detour. When they performed the task, however, they found that around one in every two people offered to go out of their way to help.

How about borrowing a stranger's mobile phone? We've surely all felt the panic of seeing the low-battery alert just before we need to make an important call, but how many of us have dared ask a passer-by if we can use theirs? This is not a trivial request; handing over your phone to a stranger comes with a small risk of theft, and the participants were given a very strict script that prevented them from using sweet talk to get what they wanted. They could pose the question, and if pressed for details, they could simply explain that they needed to 'call someone about a psychology experiment' and reassure the passer-by that they would be 'very quick', without further elaboration. Quite frankly, I'm surprised that anyone would let go of their phone for such a dubious explanation. The average participant, however, got what they wanted after asking just six people, whereas they had estimated that they'd need to try at least ten people to be able to make the call.

To test the phenomenon 'in the field', Bohns questioned volunteers taking part in sponsored marathons and triathlons for the Leukaemia and Lymphoma Society. In return for free support in their athletic training, the participants had to raise a minimum sum of between $2,100 and $5,000, depending on the event. Most of the volunteers must have felt confident that they would have been able to raise this money, otherwise they wouldn't have signed up – but they nevertheless overestimated how hard the task would be. They predicted they'd need to ask around 210 people to meet these goals, but the average participant reached their target after contacting just 122 people. And the average donation was larger than expected: around $64 compared to $48.

The 'underestimation-of-compliance effect' is extremely robust. By 2016, Bohns's research had analysed data from more than 14,000 requests – a huge sample – and the average effect size is large, with people underestimating others' willingness to help by around 48 per cent. And various other surveys show similar findings. Imagine that you are at a restaurant and have forgotten your wallet. People are much more willing to lend money to a friend in need than they would be to ask for that money themselves.[7]

Why are we so pessimistic? One likely explanation is that we pay too much attention to the trouble we might be causing the other person, and not enough attention to people's genuine concern for our happiness and desire to please. We may also underestimate people's wish to make a good impression. Put simply, the people being asked the favour want to appear kind and trusting – and denying a request might put that in peril. If they refuse to fill out a questionnaire, they might seem impatient and ungenerous; if they refuse to lend someone their phone, it will show that they are suspicious and scared of theft. For many more people than we would imagine, the inconvenience of helping – or the small risk of personal damage – is better than appearing antisocial and untrusting.

This feeling of obligation to meet social norms can sometimes lead people to commit unethical acts. In one experiment, Bohns gave participants fake library books and asked volunteers to approach strangers with the following request: 'Hi, I'm trying to play a prank on someone, but they know my handwriting. Will you just quickly write the word "pickle" on this page of this library book?' Most people predicted that it would be very hard to find someone who would comply with the strange request, but they were wrong: 64 per cent of the people that the participants approached eventually committed the small act of vandalism, despite initially raising some objections.

In a further experiment, Bohns found that people were more likely than expected to help a stranger falsify a document for a university assignment. There are obvious limits to the kinds of unethical acts you can ask people to perform in a laboratory, so Bohns has also taken to asking people to imagine their reactions to various scenarios. In these thought experiments, people revealed that a significant proportion of participants would read a colleague's private DMs or agree to buy alcohol for underage kids, if encouraged to do so by another person.[8]

If we care about others' wellbeing, this should be a warning: we must be selective about the favours that we ask, and we should listen carefully to their concerns or displays of discomfort. Authentic social connection cannot arise from persuading people to commit dubious acts they would later regret. Provided that our requests are morally sound, however, we may be surprised by the kindness of others. And in return for the inconvenience, we may be providing them with the opportunity to boost their own health and wellbeing. As we shall see, generous deeds can bring surprising benefits to the giver as well as the receiver.

WHY GENEROUS HEARTS ARE HEALTHIER HEARTS

If you are diagnosed with high blood pressure, you have many courses of action. You can try to lose weight and exercise more, cut down on your drinking, strip salt from your diet, and take up yoga or meditation. You can choose to take medications that relax your blood vessels or reduce the force of each heartbeat. Or you might decide to be a little bit more generous with your time or money.

In the mid-2010s, Canadian and American scientists recruited seventy-three people aged sixty-five and above who had been diagnosed with hypertension. Each was given three payments of $40, contained in a sealed bottle, over six weeks – along with strict instructions on how to spend it. Half were told to treat themselves, while the rest were encouraged to treat others. 'It does not matter how you spend the $40, as long as you spend it on someone else,' they were told. The participants were told to choose specific days on which to spend the money, so that a research assistant could phone them in the late afternoon to ask for details of their purchases and to check that they'd followed the instructions correctly. By the end of the six weeks, the people who had been encouraged to act generously towards others showed a significant drop in blood pressure above and beyond the effects of their existing treatments, while those who had spent the money on themselves saw no change.

The benefits were so striking that it's worth giving the precise figures. The generous group's systolic blood pressure, as the heart pushes blood into the arteries, was 113.85 mmHg, compared to 120.71 mmHg; and their diastolic blood pressure, during the rest period between beats, was 67.03 mmHg compared to 72.97 mmHg. To put that in context: this improvement in cardiovascular health is equivalent to the effects of taking hypertension medication or embarking on a new diet and exercise regime.[9]

This experiment must come with the usual caveats accompanying studies with relatively small samples of participants – and you should always follow your doctor's advice – but plenty of converging evidence points towards the same conclusion: the more generous someone's behaviour, the healthier their heart. When you question people about their monthly spending, for instance, those who give a higher proportion of their income to others are also likely to have better health, including lower blood pressure.[10] And we see similar effects with many other prosocial behaviours. People who volunteer for charities for a few hours a week are at a lower risk of developing hypertension over the subsequent years than those who spend their time on less altruistic pursuits, for instance.[11] Crucially, these differences remain when you control for a host of other factors that are known to influence people's fitness, and when you take into account any differences in people's health at the start of the study.

This is the gift of giving.[12] And according to Tristen Inagaki at San Diego State University, it may be a by-product of our hard-wired parenting skills. Inagaki's argument revolves around the vulnerability of primate young. Parents need a huge amount of empathy for their baby's needs, but they cannot afford to become overwhelmed by those emotions when they need to focus on the task of caring for the little one; otherwise, they may simply recoil, or feel too uncomfortable to provide the necessary care. Inagaki argues that our ancestors' brains may therefore have evolved to suppress their stress response while caring for their young. The brain would also have evolved to create feelings of reward once help had been delivered – to cement the bond between parent and child and to ensure that the adult continued to engage in the caring behaviour again and again. While the stress-suppression-and-reward response may have first evolved to improve the survival of helpless babies, over time it may have started to kick into action whenever we care for any other being. In prehistory, this would have helped our ancestors to act more cooperatively, resulting in our increasingly large social groups.

A slightly sadistic experiment supports Inagaki's hypothesis. She invited romantic couples into the lab and administered painful electric shocks to one member of each pair, while the other attempted to comfort them. Brain scans of the person offering support revealed heightened activity in some of the regions known to underlie parental care, including the ventral striatum, which is associated with reward, and the septal area, which is thought to control stress responses. The increased activity in the septal area was accompanied by decreased activity in a separate region known as the amygdala, which processes our emotions and is typically more active when we feel fearful and anxious. Inagaki found similar results when looking at people's responses to financial generosity, but there was a catch. While providing money to a specific person resulted in stress suppression, this was not true of altruism offered to a charitable organisation. The automatic psychological and physiological response to giving seems to arise from our care for individuals, rather than abstract causes.[13]

By soothing our stress response, generous behaviour may not only lower our blood pressure. It can also slow the release of the stress hormone cortisol, lower our cholesterol, soothe the symptoms of chronic pain and dampen down the inflammation that can result in wear and tear on our tissues.[14] There are even some signs that generosity could reduce the speed of our cellular ageing. As the years go by, the activity of our genes tends to change in a characteristic way, and those changes are associated with a greater risk of disease. Scientists call this the 'epigenetic clock', and for people who regularly help others it appears to tick more slowly.[15]

The long-term health benefits can be seen in a five-year study of 846 people in Detroit, Michigan. At the start of the study, the participants were asked to list any stressful events – such as being burgled, losing their job, or suffering a bereavement – that had taken place in the past year. They also declared how often they helped others by giving lifts, for example, or regularly running errands, providing childcare, or doing the shopping for someone less mobile. The researchers then kept track of who lived and who died over the following half a decade. Confirming the well-known effects of stress on our health, the citizens who had experienced a disruptive event were around 30 per cent more likely to die over the study period. This was not the case, however, for those who engaged in the highest levels of helping; their generous behaviour had eliminated the link between their life stresses and premature death.[16]

This is not to mention the psychological benefits. Providing social support improves our general life satisfaction and makes life more meaningful, with studies from countries as diverse as Canada, Uganda and India demonstrating that generosity is one of the best routes to personal happiness.[17] Such findings need to be taken in perspective: I'm certainly not arguing that we should *always* put others' needs above our own. When we feel close to breaking point, we must prioritise our own wellbeing. But on those days when we feel lonely, irritated and dissatisfied with our lives, an altruistic act may just be the best form of self-care.

THE BENJAMIN FRANKLIN EFFECT, AKA *AMAE*

Besides encouraging us all to be a little more generous ourselves, the gift of giving should relieve our fears of asking for assistance from others. Provided that our requests are sensitive and respectful, many people will be genuinely happy to help, and afterwards they will bask in the warm glow of their altruism.[18] Given these good feelings, there should be little wonder that requesting assistance can work to enhance social connection. Benjamin Franklin, of course, proposed as much in his autobiography. It was only in the 1960s, though, that psychologists offered experimental proof for this idea – which forms our ninth law of connection.

In an eye-catching study, David Landy and Jon Jecker first invited participants, one by one, to play a game for money – with winnings of either 60 cents or $3. (Remember this was the 1960s; the sums would be much bigger today.) Landy and Jecker had told the person conducting the experiment – Mr Boyd – to be brusque and cold throughout the process, and as the participants left the lab, they experienced one of three scenarios. In the first, the players were told they could keep their winnings. In the second, the participants were asked to return their money as a donation to the university psychology department. And in the third, Boyd asked if they would do him a favour and give *him* the winnings. 'The funds for the experiment have run out and I'm using my own money to finish the experiment,' he told them. 'As a favour to me, would you return the money you won?' Finally, they were asked to rate their liking of Boyd and whether they would be happy to help him with more research.

The participants in the first condition did not like Mr Boyd at all. They gave him a rating of just 5.8 out of 12 – an overall negative assessment. The participants who had given him their winnings, however, tended to feel much more warmly towards him – and the greater the winnings they donated, the more positive they felt. Those who had donated $3, for example, gave Boyd a rating of 7.6, suggesting that the simple act of generosity had helped them to overcome their initial bad impression. This was not true if they'd simply donated the money to the department – it was the act of helping Boyd, in particular, that increased their feelings of warmth towards him.[19]

Landy and Jecker's findings have received relatively little attention over the decades, but further evidence for the ninth law of connection comes from recent research of Yu Niiya at Hosei University. She has studied the Japanese concept of *amae*, which describes someone's tendency 'to depend and presume upon another's love or bask in another's indulgence'. English writers have translated *amae* as 'coaxing', 'wheedling' or 'being spoiled and pampered' – all of which have distinctively negative connotations. But according to Niiya's research, *amae* may be a secret for happiness in many relationships.

In her view, an appeal for support can be characterised as an '*amae* request' if it concerns a task that the requester is perfectly capable of doing on his or her own, and if the fulfilment of the task will underline the bond between the two people. That may sound a little childish, and it is often associated with infant behaviour: a six-year-old might pester their mum or dad to read them a story, not because they are incapable of reading it themselves but because they want to feel cared for and looked after. But adults will also engage in this all the time: asking your partner to buy you a gift even if you have enough money to pay for it yourself, suggesting that your friend picks you up from the airport even though you can take public transport, or persuading your partner to cook your favourite meal even though they might be busy with other tasks would all be considered as *amae* requests, according to Niiya.

In her research, Niiya was interested to see how the people on the receiving end of *amae* requests would perceive the pleas for help and support. Would the inconvenience generate anger and resentment? Or would it lead the person to feel needed, valued and respected? And would this differ between cultures? To find out, she first created vignettes that represented *amae* requests, such as:

> Your best friend S who is also your roommate has to write up a term paper by tomorrow morning and is burning the midnight oil. You decide to go to bed before S because you have an early class. At midnight, S's computer breaks down. It would take the whole night if S tried to fix it alone. S has several friends who are good at computers, and you are one of them. S knows that you can fix the computer very easily.
>
> On a scale of 1 (not at all) to 7 (very much), how happy, sad, irritated and disappointed would you be if they asked you to fix the computer for them? And how close to them would you feel after they made that request?

Niiya compared responses to this story to those of people who read the same vignettes, but discovered that S had struggled on their own, or that S had asked another friend – not you – to do the favour.

The first experiment examined Japanese students. As you might expect, the participants imagined that they'd be a little irritated to have been woken at midnight, but they still predicted feeling far happier to have been asked to help out themselves, rather than hearing that S had turned to a tech company or another friend to solve their problems – by quite some margin. Those who imagined receiving the request reported 4.5 on the seven-point scale, compared to 2.9 and 3.4 in the other two conditions. And those warm feelings were reflected in the ratings of closeness. The participants expected to feel closer to S if they had been asked to help out, despite (or perhaps because) of the inconvenience. Niiya replicated these results with a host of other requests – such as being asked to dog-sit while the owner takes part in an out-of-town training course, or being asked to put someone up in their flat for three days. In each case, the requests increased the perceived closeness of the relationship.

Niiya's next experiment aimed to find out if people outside Japan would react in the same way, and the answer was yes. She found that students at the University of Michigan generally responded very positively to the descriptions of the requests; they imagined feeling closer to the person who had asked for help and saw it as a sign of stronger friendship.[20]

Moving beyond these hypothetical scenarios, Niiya has since tested the effects of *amae* in laboratory experiments, in which participants were assigned a partner and asked to complete three brain-teasing puzzles. Unknown to the participants, the set-up was rigged, so that they always received easier puzzles, and their 'partners' were actors who had been primed to ask for help as soon as the real participants had finished their set. 'Wow, already? That was fast! Would you mind helping me with this?' the actors exclaimed – and passed over one of their puzzles. Both before and after the task, the participants were given questionnaires in which they rated how sociable their partner appeared, and how much they liked them. Sure enough, both ratings increased after the partner had asked for help. Crucially, this did not happen when the *experimenter* asked the participant to help their partner, instead of the partner making the request themselves. For *amae* to increase affection, it seems that the request must come from the same person who is going to receive the support; you have to humble yourself in front of them.[21]

Why does giving – as well as receiving – social support strengthen our connections in this way? The classic explanation, given by Landy and Jecker in that paper from the 1960s, is that these requests provoke 'cognitive dissonance'. After someone has given you support, they struggle to explain how they could have behaved so kindly to someone who was not also a close friend. To resolve that inconsistency, they update their beliefs about you to be more positive. You can imagine the retroactive thought process – 'I must have really liked S if I were willing to get up in the middle of the night to fix their computer.'

The increase in affection could also arise from the humility on display. Your request makes you seem less threatening, and the fact that you asked them – of all people – is a sign of trust in their capability to help you. Like other forms of self-disclosure, the admission of need and vulnerability has provided an insight into your mind, which reinforces the sense of shared reality. Attempting to manage your task by yourself might have shown greater independence but it would have blocked all those opportunities for connection.

Or perhaps it is because *amae* appeals to people's parental instincts. Remember that acts of generosity can trigger the same emotional responses that first evolved to help our primate ancestors look after their young, including stress suppression and reward. *Amae* requests may capitalise on the same process. They bring pleasure to the person receiving and the person giving the support, which should produce a greater sense of connection on both sides.[22]

Whatever the mechanism behind *amae*, a recognition of its power could make our lives far easier. Niiya has shown that people can apply it in the workplace, for example. Questioning Japanese entrepreneurs and managers about their relationships with their co-workers, she found that the more someone recognised the power of *amae*, the more likely they were to seek help and advice from the people around them, and the less worried they were about the potential consequences of those requests. They also reported greater trust and engagement. They gave higher ratings to the statement 'I am capable of becoming fast friends with anyone', for example. Along similar lines, Niiya has shown that first-year undergraduates who employ *amae* tend to adjust more quickly to the university environment, with greater feelings of purpose and life satisfaction at the end of their first year.[23] *Amae* can also improve relationship satisfaction. One survey questioned Japanese heterosexual couples about their interactions each day for two weeks, and found that the more they engaged in *amae* with each other, the happier they were together.[24]

Reading about *amae*, I couldn't help but recall a beautiful observation from the psychologist Alison Gopnik in her book *The Philosophical Baby*. 'It's not so much that we care for children because we love them,' she wrote, 'as that we love them because we care for them.'[25] Gopnik was writing about the perspective of a parent, but it appears to reflect a truth that governs many other relationships.

KNOWING WHEN AND HOW TO ASK

Requesting help will always require great discretion. We are unlikely to improve our relationships by becoming a needy nuisance, or heavy-handedly persuading others to help us when they are already buckling under the pressure of their own lives. For this reason, it is essential that we offer plenty of opportunities for people to decline our requests without losing face.

One option is to provide verbal recognition of the other responsibilities the person is facing. I am lucky to be part of a network of mutually supportive writers, who help to brainstorm ideas, read drafts and soothe my frustration when the words don't flow as I would like. Whatever the favour I'm asking, I always acknowledge the sheer number of similar requests my friends may be receiving on top of their regular work, and to apologise if I am not contacting them at a convenient time. More often than not, I am surprised by just how quickly these acquaintances will offer to help, despite their busy schedules – but I hope that my wording makes it clear that I would fully understand a refusal without any offence being taken.

In the workplace, we should be especially conscious of the power dynamics at play. As Vanessa Bohns describes in her book *You Have More Influence Than You Think*, people occupying higher positions of power tend to spend less effort thinking about others' perspectives.

The reason is practical: when you're top of the pecking order, you simply don't need to spend so much time thinking about the motives and feelings of the people around you. As a result, a boss may be inclined to underestimate how difficult it will be for one of their team to take on a new responsibility, and they will forget how hard it can be to say no.[26] When in doubt, follow Seneca's rule: 'Treat your inferior as you would be treated by your betters.'[27]

If we are asking a favour that will require significant investment on the other person's part, it is often best to give them some time to mull over our request, so that they do not commit to something they will later regret. We might ask by email, for instance, rather than placing them on the spot in a face-to-face meeting or video call, since it will be easier for them to pause and compose their thoughts before accepting or declining. Bohns's research shows that people often forget how much harder it is to decline a request for help in a face-to-face conversation compared to an email, but the difference is substantial.[28] However we ask, we should be clear that we won't expect an instant answer, and we should make sure that we are upfront about any inconveniences that the favour might incur. While we may be more likely to get the response we want by sugar-coating the difficulties involved and demanding an immediate reply, the other person will end up resenting those tricks.

Once someone has accepted our requests for support, it is essential to show our gratitude for their assistance. Whether they have simply warmed our mood or enabled us to achieve something that we would have otherwise been powerless to do, we need to let them know the impact they have had on us. This is important for all the reasons outlined in Chapter 6, and more: experimental studies of altruism show that the psychological and physiological effects are most pronounced when the generous behaviour is seen to have been effective at improving the circumstances of the other person. This is evident in brain scans: the more a carer feels effective and appreciated, the more likely they are to experience that mix of stress suppression and reward that often accompanies altruism. By taking others' help for granted and failing to let them know the difference they have made, we will deprive them of those benefits.[29]

Finally, we should ensure that our relationships are reciprocal. Given the underestimation-of-compliance effect, we cannot simply assume that the people close to us will let us know when they need our support, and so we should be a little more proactive in offering a helping hand. In our conversation, we can remind the people around us that we are available for support whenever they need it. Simply asking how someone is feeling may allow them to open up about a problem they'd previously kept to themselves, but we can also be explicit: 'What can I do to help you today?' It's the simplest of questions that is often forgotten in everyday exchanges, but could do so much to create a more supportive social network.

As Benjamin Franklin recognised in his early political career, social connection involves giving *and* receiving, and with a little effort we can all bask in the pleasure of mutual generosity.

What you need to know

- We are too pessimistic about others' willingness to offer support. In general, people are about twice as likely to help us as we expect, and they are more likely to have enjoyed helping us out than we might predict
- When someone acts generously, it creates a sense of reward and soothes their stress response. In the long term, this can bring numerous health benefits, including reduced blood pressure and slower cellular ageing. This is the 'gift of giving'
- People like us *more* after we've asked them for a favour – even if it creates inconvenience. This is known as the Benjamin Franklin effect. In Japanese, this phenomenon is described as *amae*

Action points

- If you find that someone is struggling to warm to you, consider asking them for a small favour. Thanks to *amae*, this may change their opinion and make your future interactions easier

- Do not fear making requests, but be sure to give the other person plenty of opportunities to refuse. Favours provided through obligation rather than generosity are unlikely to improve your relationships
- If you can afford it, set aside a small amount of money each month to spend on the people around you. Research suggests that treating others will make you happier and healthier than treating yourself
- If you have enough time, consider volunteering for a cause you care about. Partly because of their increased social connection, volunteers tend to enjoy better health

CHAPTER 10

HEALING BAD FEELINGS

If you have ever struggled to find the right words to express your sympathy for a friend in need, you may find some courage in the fact that Henry James – one of the English language's greatest stylists – felt the same.

In July 1883, James received a series of heart-wrenching letters from his friend Grace Norton, who had plunged into an all-encompassing depression. He began his reply with a humble recognition of his ineloquence: 'My dear Grace, Before the sufferings of others I am always utterly powerless, and the letter you gave me reveals such depths of suffering that I hardly know what to say to you.' But he then went on to write, with the 'voice of stoicism', a wise meditation on the nature of sadness and our abilities to endure emotional pain.

> Sorrow comes in great waves – no one can know that better than you – but it rolls over us, and though it may almost smother us it leaves us on the spot and we know that if it is strong we are stronger, inasmuch as it passes and we remain. It wears us, uses us, but we wear it and use it in return; and it is blind, whereas we after a manner see. My dear Grace, you are passing through a darkness in which I myself in my ignorance see nothing but that you have been made wretchedly ill by it; but it is only a darkness, it is not an end, or the end . . . Everything will pass, and serenity and accepted mysteries and disillusionments, and the tenderness of a few good people, and new opportunities and ever so much of life, in a word, will remain.[1]

More than 100 years later, through technological media that James could not have begun to imagine, his letter continues to comfort modern readers facing twenty-first-century struggles; I count myself among them.

We've already explored the psychology of giving and receiving support, but emotional care requires particular attention and sensitivity. Whether someone is facing a single bad day or dealing with impossibly sad news, we may have every good intention of offering our sympathy, while doubting our abilities to help. When should we listen passively, without judgement, and when should we try to shift their perspective on this issue? Is tough love ever justified?

Each decision seems fraught with the danger of making the situation worse – but our new knowledge of shared reality and social connection inspires some much-needed guidance. With these principles in mind, we shall begin to understand why James's wise response to Grace Norton gives such consolation to so many – and how we can do the same.

THE POWER OF SAYING 'I'M HERE'

First, the good news: most of us are much better at comforting others than we believe, and our support is more valued than we expect.

In one survey by James Dungan and colleagues at the University of Chicago, students were asked to think of someone on campus who had been going through a tough time and might appreciate a message of support. On a scale from 0 (not at all) to 10 (extremely) they then estimated how that message might be received by that person, with separate ratings for feelings of awkwardness, perceived warmth and social competence – that is, whether they had chosen the 'right' words to express their sentiments. Finally, they sent their letter, along with a link to a survey measuring the recipients' interpretations of the letters, using the same criteria.

In line with the other research on misguided social appraisals, many participants were highly pessimistic about their capacity to support the other person, but these fears were not grounded in reality. On average, their letters were considered significantly warmer and more articulate than they had expected, and most of the people receiving the notes felt comforted by the sentiments behind the words. Contrary to the participants' initial beliefs, the nature of the relationship did not influence how the letters were interpreted. Whether the person was a very close friend or a more distant acquaintance, the kind deed was equally appreciated in each case.

To find out whether this would also be true of face-to-face conversation, a follow-up experiment paired participants with random strangers who had previously described a difficult situation that they were facing, such as financial problems, romantic woes, family disputes and illness. The participants' job, in a subsequent fifteen-minute conversation, was to provide as much social support as they could on the issue in question. 'You could express empathy, give advice, share something from your own life, offer some kind of assistance, or anything else. We only ask that you try to express your support in whatever way makes sense to you.' Once again, the participants' expectations of those conversations were completely skewed. They worried that they would be unable to find the right things to say and they expected the chat itself to be uncomfortable and draining. Yet both sides found it to be an affirming experience.[2]

THE PERILS OF VENTING

Just because our overtures of help are generally well received does not mean that we cannot improve the ways that we handle these tricky situations. When we offer emotional support to others, many of us are rather passive: we 'lend an ear' as they vent their feelings, under the belief that the mere expression of their troubles will do them good. This idea has its roots in an old theory known as the 'hydraulic model of emotion'. The theory took inspiration from the pipes and pistons that would come to revolutionise manufacturing, as Enlightenment thinkers began to view the brain as a kind of pneumatic machine. In much the same way that an unfortunate obstruction may lead an engine to explode, the free flow of 'spirits' around the brain was thought to be essential for a sound state of mind.

Later scientists came to see this as a metaphor rather than a physical description of the way the brain worked, but the general idea stuck, and in the late nineteenth and early twentieth century Sigmund Freud proposed an explicit link between emotional blockages and mental illnesses such as 'hysteria'. Freud's solution was psychoanalysis, but the hydraulic model is evident in the ways that many of us choose to process our feelings in our everyday lives.[3] It is embedded in our language: we encourage people to 'get it off your chest' or 'let off steam'. According to surveys from Europe and Asia, around 80 per cent of people believe that we – or the people we care about – can release negative feelings by sharing them with others.[4]

There is good reason to think that this would help. As the research on secrets showed us (see Chapter 7), hiding important elements of our identity can feel like a physical burden that eases after we have told other people – and we may hope to provide similar relief when someone bares their pain to us. Unfortunately, recent research suggests that the feeling of release is fleeting, while the long-term benefits of venting are severely limited. And if it is repeated too often, unrestricted emotional disclosure can be actively damaging.[5]

Consider studies examining the aftermath of tragedies such as the Virginia Tech school shootings and the 9/11 terrorist attacks. If the hydraulic model were correct, you might expect that people who regularly discuss their feelings with others would have the best psychological outcomes – compared to those who bottle up their feelings and keep their fears to themselves. Yet researchers have found that the opposite is often the case: people who spent the most time expressing their sadness and anxiety were no less likely to develop symptoms of depression and anxiety disorders, and in some cases it actually seemed to increase the risk of PTSD.[6] One reason is that we keep on reactivating the same neural circuits associated with the hurt; rather than releasing the bad feelings, repeatedly rehashing upsetting events leads them to become more deeply embedded in our minds, so that they occupy our thoughts for longer and longer.

Psychologists often describe this habit as 'co-brooding', and it can be harmful in many different kinds of relationships. We can see this in the angst-ridden exchanges of teenagers, for instance, who may spend hours describing their woes to each other with seemingly no potential for closure. While those conversations may foment a sense of connection – experiencing and articulating the same stresses is a form of shared reality – they can amplify distress if they become too frequent.[7] The same goes for adult couples. There is an undoubted allure to having your soulmate share your burdens, but the more two people vent their feelings to each other, the unhappier they both feel in the long run.[8] At the physiological level, brooding conversations increase levels of the stress hormone cortisol, which could cause bodily harm if it spikes too frequently.[9]

Such findings may seem dispiriting for anyone who values emotional disclosure, and few psychologists would encourage people to suppress their feelings. It really is good to talk. But the healthiest conversations will help the person to move beyond their immediate pain, by encouraging greater wisdom and insight into their problems. And to do that, we need to engage in 'co-reflection'.

One of the best examples of this subtle but essential distinction comes from studies of people's communication during the early stages of the Covid-19 pandemic. When the world seemed to be falling apart around us, it was natural that our conversations would take a darker turn. For some people, however, the chat was almost exclusively focused on expressing negative emotions; they endorsed the statement: 'we talk a lot about all of the different bad things that might happen because of the pandemic'. As the previous research on co-brooding would have predicted, these people were considerably more likely to suffer from worse mental health than people who talked less about the pandemic.

Others, however, took a more philosophical stance. They endorsed the statement 'we spend a lot of time trying to figure out parts of the Covid-19 situation that we can't understand'. Such conversations would have helped to reduce the uncertainty and motivated them to take small steps to improve their situation, such as looking for practical means to cope with home-schooling their children or looking for ways to donate face masks for the vulnerable. As a result, this conversational approach – co-reflection – was associated with better mental health.[10]

Note that neither conversational style requires a Pollyanna-ish 'glass half-full' attitude to life; co-reflection can still acknowledge the strains we are under, while simultaneously attempting to gain a little more perspective and make sense of the situation at hand.

We can see a similar phenomenon in studies of 'expressive writing'. As the name suggests, this intervention involves committing our deepest thoughts and feelings to paper – and initial research suggested that this was extremely good for people's mental health. The benefits can be variable, however, and later linguistic analyses of people's essays helped to identify the features that predicted greater emotional recovery. Whereas some people simply articulated the pain and hurt they'd been suffering – which, to be fair, was what the instructions had asked them to do – others took a more reflective and philosophical stance. This was marked by a greater preponderance of words examining cause and effect (such as 'because' or 'therefore'), insight ('think', 'believe', 'consider' or 'understand') and inhibition ('stuck', 'forced' or 'compelled'). The choice of these words suggests that the writers were thinking more reflectively about their issues and drawing new conclusions about the effects on their lives, which in turn contributed to better mental health in the months after the intervention.[11]

When someone is distressed and immersed in the waves of bad feeling that threaten to engulf them, it will be incredibly difficult for them to think about their problems in these more abstract terms. And it is at precisely these moments that they will need a loved one to gently encourage a change of perspective.

AUTOBIOGRAPHICAL REASONING AND EMOTION COACHING

The emotional tenor of our conversations may be particularly important in a child's early years. Indeed, family dialogues from decades past may still be influencing your mental health today. To understand why, we need a quick primer on the development of our autobiographical memories. In the first few years of life, most children can remember only the slimmest shards of their experiences – the feel of sand on the beach, the prick of a needle in a doctor's surgery. These may get more detailed as the child learns more and more vocabulary, but they are largely disconnected from each other; they remain isolated sketches of single events. It is only after years of development that the child can slot their recollections into a narrative that has a coherent structure. By the end of adolescence, that narrative may adopt the form of a novel. The teen will start to recognise key events as turning points, with new 'chapters' that represent new eras. The psychologist Dan McAdams at Northwestern University in Illinois describes this as the transition from 'actor' to 'author'.[12]

Thinking about our life's trajectory in these terms is known as 'autobiographical reasoning', and it provides one important means of gaining perspective on our current failures, fears and frustrations. We may still feel pain from a break-up but we can also recognise that, having left someone who was not treating us fairly, we have learned important lessons about our self-respect, for example.[13] Or we may be facing a particularly stressful project at work and recognise that we've already shown great resilience and determination in the past, perhaps during our time at university – giving us the strength to deal with our current challenge. For these reasons, the quality of someone's personal narrative is often linked to their mental health. People with more coherent and detailed stories, who are better able to extract meaning from their memories, are less susceptible to depression, for instance.[14] They are also more likely to report better life satisfaction and a greater sense of purpose in their lives.[15]

We picked up these skills from our conversations with our caregivers – parents, grandparents, uncles, aunts, elder siblings or family friends. And research from Elaine Reese at the University of Otago in New Zealand suggests that some people are much better tutors than others, with lasting consequences. When caregivers validate a child's memories and ask them to elaborate on the details, that child is more likely to grow up with a richer autobiographical memory, Reese has shown. As a result, they will find it easier to extract meaning from stressful events and to place them in the broader context of their lives, which contributes to better wellbeing. If the caregiver ignores this cue or brushes it away, however, then the development of the child's autobiographical memory will be stunted – reducing their overall emotional resilience later in life.[16]

Many adult–child conversations will focus on a child's feelings about an event and how to deal with them – and these too can be crucial for the ways we form memories of distressing events. Researchers in the UK recruited 132 families who had recently undergone a traumatic event, such as a house fire or car accident, which had resulted in a hospital visit. Around one month after the trauma, the researchers recorded the family having a detailed discussion about the trauma, in which they recalled the way the event unfolded, and described their thoughts and feelings about what had happened.

The contents of those conversations varied dramatically from family to family. Some parents tended to focus on the worst-case scenarios, making statements like 'You could have been killed'. Or they catastrophised the consequences – 'We'll never be the same again' – and encouraged fear-related avoidance; if an accident happened on a particular road, then they vowed never to drive the same route again. Others were more nuanced. They didn't shy away from the fear and upset the child was feeling, but they also praised the child for how well they had coped, for example, and emphasised their resilience in the future – 'You'll be back to your usual self once your leg is better'. And these characteristics could predict the child's wellbeing six months after the event, including their symptoms of post-traumatic stress – even after the researchers had controlled for the initial level of the child's trauma.[17]

More constructive conversations of this kind are sometimes known as 'emotion coaching', and we can see similar effects on children undergoing surgery. It is natural to talk about their anxiety, fear and discomfort, as it helps them to cope with the operation and make meaning of the traumatic event – but some parents take it too far. In their concern for their child's welfare, they focus too much on the pain, which leads the child to exaggerate the trauma they experienced.[18] This will only increase their fear of hospitals in the future.

Effective emotion coaching helps to balance recollections of discomfort with more positive features of the event – such as the child's resilience. If a child remembers how much they cried, for example, the parent might try to remind them how quickly they regained their composure, or how they managed their pain with deep breathing. When families engage in this kind of conversation, the children tend to have more nuanced memories of their operations and they have less fear of doctors and hospitals further down the line.

Such changes occur in the neurological fabric of the child's memory. Each time we recall something, the brain reactivates the neural networks storing the event, so that it becomes malleable, before it is reconsolidated. If our conversations about an event focus only on the most negative and frightening aspects, those elements are going to be strengthened within the memory trace, but if the conversation helps us to find more positive meanings, then we will continue to recall the more nuanced interpretation in the future.

Some caregivers will be natural at emotion coaching, whereas others might need some guidance on the best ways to provide it for their children – particularly if their own parents had failed to offer such positive reappraisals themselves. Fortunately, recent research suggests that these skills can be trained. Just a few guidelines on the best ways to talk to children about trauma can help caregivers provide more constructive conversations with their children.[19]

It is never too late to learn these skills. Whether we are talking to a friend, colleague or family member, our conversation can provide the necessary shift in perspective that is essential for healing to begin.

GETTING CLOSE AND GAINING DISTANCE

If we are to put these principles into practice, our offers of emotional support should comprise two distinct steps.

The first is validation. People crave mutual understanding at the best of times, and our desire for the most fundamental essence of social connection will be particularly keen during emotional distress. To provide validation, we can reiterate what they have said to demonstrate attentive listening, ask them to elaborate on the ways they are feeling, and affirm the fact that we consider those emotions to be legitimate. This does not mean that you have to agree with everything they say – your reactions to a similar situation may be completely different – but you can show that you acknowledge their point of view and that you are doing your very best to comprehend the situation at hand.

During this stage of the conversation, we should make sure that we remain humble and avoid jumping to conclusions before the person has fully expressed what they are feeling. It is easy to hear only one part of the story before we start offering our thoughts, when we've completely misunderstood the crux of the issue. This will only create a sense of alienation – even if your intentions were well-meaning. And even if we do have questions about the appropriateness of their reactions, we must try not to use judgemental language, which is only going to add to their upset, and will make it considerably harder for them to find useful insights or solutions to their problems.

Once we have reasserted a shared reality, it is time for us to shift the conversation to reflection and reappraisal *before* it descends into brooding. I see this as providing the 'voice of stoicism' that Henry James adopted with Grace Norton; it attempts to place a slight distance between the person and their distress. We might gently encourage them to look for alternative explanations for an upsetting event, for instance, rather than jumping to the worst-case scenario.

To take a trivial example, imagine that your partner is upset that someone close to them has forgotten their birthday. You could just tell them to stop being so oversensitive, but that might leave them feeling foolish and ashamed of their very real feelings. Another response might be to take their side and to feel outraged on their behalf, a behaviour that may leave them feeling validated, but which may also amplify their worst-case thinking – that this other person did not care about them. The better solution would be to empathise with their worries and concerns, and to offer some alternative explanations for the friend's behaviour. Perhaps the friend was going through a difficult period at work, or one of his children was sick. You might also remind them of all the other people who *did* get in touch, so that they can put this disappointment in the broader perspective of their wider social network, so that the negligent behaviour of one friend does not seem so catastrophic.

In other instances, we might encourage the 'autobiographical reasoning' that could help a friend to see their current predicament as a turning point in their life story. Or we might remind them of their resilience and resources, and help them to recognise a previous period of their life in which they had overcome enormous hurdles, and encourage them to think of practical solutions to their problems.[20]

In each case, we must proceed tentatively. We cannot arrogantly tell someone how to think of their situation; we can only offer some suggestions. Sometimes it may be enough to ask the right questions and allow the other person to come to the new perspective by themselves, as Ethan Kross at the University of Michigan, Ann Arbor, and colleagues have demonstrated. Kross is a leading expert on the differences between rumination and adaptive self-reflection, and he has developed cognitive techniques to find new perspectives on upsetting events. For example, he has found that it can help to imagine viewing our situation at a later date, such as ten years in the future – a strategy that helps to shrink the emotional significance of the event, so that it no longer feels all-consuming.[21]

Working with a large team of psychologists, Kross examined whether it would be useful to incorporate this kind of prompt in our conversations. To do so, they recruited nearly 200 students who were in the midst of an emotional upset, and asked them to discuss their situation, over online direct messages, with a trained experimenter who was – at that stage – ignorant of the study's aims or hypotheses. In some of the trials, the experimenter simply asked the participants to *recount* their experiences.

1. Can you tell me about what happened – what happened and what did you feel – from start to finish?
2. What went through your mind during the exact moment?
3. What stuck out the most at that moment?
4. What did (he/she/they) say and do?
5. How did this make you feel at that moment?

Such questions demonstrate curiosity, but they focus on the concrete details – *what* happened to the person and *how* they felt. For other trials, the experimenter encouraged the participants to *reconstrue* the event using the following questions:

1. Looking at the situation, could you tell me why this event was stressful to you?
2. Why do you think you reacted that way?
3. Why do you think the other person reacted that way?
4. Have you learned anything from this experience, and if so, would you mind sharing it with me?

5. In the grand scheme of things, if you look at the 'big picture', does that help you make sense of this experience? Why or why not?

The series of *why* questions – as opposed to *what* questions – immediately encourages greater causal insight. The last two points then persuade the person to look at the event through a much wider lens, so that they can find greater meaning in what happened.

The effects of the two different strategies were immediate and revealed just how important it can be to encourage a change in perspective. Contrary to the hydraulic model of emotional expression, the participants who simply recounted the event ended up feeling considerably more miserable after the conversation than they had at the start; the events were still raw. Those who had reconstrued the events, in contrast, had already started to feel greater closure. The conversation was helping them to put their feelings to rest.[22]

We may already apply some of these principles, on occasion, but social-psychological research suggests that we do not do so nearly as often as we could.[23] We are so focused on validating people's feelings and meeting their immediate emotional needs that we forget to encourage this perspective shift, or we may worry that our attempts to encourage reappraisal could appear to be dismissing others' pain or discomfort. These are reasonable concerns, and I admire Kross's research because it suggests a very tactful way of encouraging reappraisal. Without pushing people towards a new point of view, we can try to provide the necessary space in a conversation for those thoughts to emerge on their own.

LETTING GO

One final guideline for providing emotional support is both simple to understand and difficult to apply. And that is to know when to relinquish control, and how to accept it.

You may be familiar with the scenario. Your partner, friend or family member seems tired, withdrawn or unusually irritable. They lack enthusiasm for their usual pursuits, or snap when they would typically be patient. You know that something must be upsetting them. When you probe their feelings, however, they deny there is a problem. Such a lack of disclosure can feel extremely hurtful: it can lead us to feel that they do not trust our judgement, and that they are putting up a barrier in our shared reality. But we should not let frustration get the better of us. The worst thing anyone can do is to use ultimatums or false dichotomies – 'I'm not speaking to you until you tell me what's wrong' or 'if you cared about me you'd open up' – to get the other person to talk. People may crave social support but they also value independence, and efforts to control someone's behaviour will only make it more likely that they'll clam up.

Researchers at Université de Montréal and Université Laval in Canada asked 268 participants to consider the different potential strategies their partners might use to show support during a difficult time, and the most popular all allowed them to retain their autonomy. The researchers found very similar patterns in a lab experiment. They invited couples into the lab and asked one partner to initiate a conversation about an important subject – such as whether or not they should decide to have children – that they had not yet resolved together, which the researchers filmed and then analysed. As you might expect, some partners quickly clam up when they approach such topics – and attempts to pressure them into speech only made them more reluctant to talk. Instead, it was the people who created conversational space, and allowed their partners to speak when they felt ready to open up, who tended to have more constructive exchanges and greater relationship satisfaction.[24]

Based on these findings, the researchers suggest that 'autonomy-supportive communication' involves three characteristics:

- Noticing a change in someone's emotion
- Expressing our concern
- Offering flexibility in the timing, topic and depth of the conversation

In real life, this might be as simple as saying 'I can see you're upset, and I'm available to talk whenever you need me', before backing off and giving them the space they require. No matter how good our intentions to support someone, we must respect their desire for solitude. There is a time and a place for engaged conversation, and that has to be at our loved one's choosing.

THE PARADOX OF COMPASSION

Is it ever reasonable to pull away from those on the brink of a crisis? Talking through someone's anger, disappointment or fear can be exhausting, after all. In our individualistic society that prioritises personal wellbeing, you may wonder whether it is really so bad to back away and allow others to deal with their problems by themselves. No one wants to be considered a fair-weather friend, but we might choose to protect ourselves from bad emotional energy.

If you have these doubts, you are not alone: scientists have taken this question very seriously. They measure 'compassion for others' (as opposed to self-compassion) by asking people to rate, on a scale of 1 (almost never) to 5 (almost always), statements such as:

- If I see someone going through a difficult time, I try to be caring towards that person
- I pay careful attention when other people talk to me
- Everyone feels down sometimes: it is part of being human

And:

- Sometimes when people talk about their problems, I feel like I don't care
- When people cry in front of me, I often don't feel anything at all
- I don't feel emotionally connected to people in pain

You will have probably guessed that the people who give high scores for the first group and low scores for the second are considered to have more compassion for others.

The third item – 'Everyone feels down sometimes: it is part of being human' – deserves particular attention. This element is sometimes called 'common humanity' and it prevents us from blaming others for their bad feelings, or feeling superior and judgemental towards people who are facing problems that are different from our own.

Researchers have compared these scores with various measures of life satisfaction. If increased empathy and tenderness for others created a personal burden, you would expect those with higher compassion to show worse emotional wellbeing – yet the very opposite is true. More compassionate people are considerably happier than those who isolate themselves from others' pain.

This is known as the 'paradox of compassion'. One good explanation is the gift of giving that we explored in the previous chapter. When we act kindly to another person, we feel a warm buzz that calms our stresses, both physiologically and psychologically – the ancient, evolved response to caregiving that increased humans' parenting skills. This may certainly play a part. Another possibility is that we see our acts of kindness as a kind of transaction or investment: having supported someone else in their hour of need, we can expect them to do the same for us, which leaves us feeling more secure.

If this idea of tit-for-tat sounds a tad cynical to you, I empathise: it's not the way I would like to view my own relationships. To me, a much likelier explanation is that compassionate people may simply be better at keeping strong friendships. They're not necessarily expecting a direct 'return on investment' for a single action, but are simply behaving in a way that encourages social connection more generally. Along these lines, researchers in Houston and Arizona measured people's compassion and compared those scores with a questionnaire on 'friendship maintenance' behaviours, which included:

- Reminiscing about shared experiences
- Sharing your private thoughts
- Making time for others even when you are busy
- Repairing misunderstandings
- Listening without making any judgement

These skills would not be useful only in times of crisis; they are essential for creating a shared reality with someone at any point in a developing relationship. For these reasons, friendship maintenance behaviours are a strong predictor of the quality, commitment and satisfaction of the relationship. The team's statistical analyses suggested that these behaviours 'mediate' the association between compassion and happiness, resolving the apparent paradox. Put simply, the more compassionate you are, the better you are at maintaining friendships, and the greater your life satisfaction resulting from those stronger connections.[25]

Such benefits were evident during the Covid-19 pandemic. In the spring of 2020, a multinational team led by Marcela Matos at the University of Coimbra in Portugal and Paul Gilbert at the University of Derby in the UK recruited 4,000 people from twenty-one countries, stretching from Brazil to Australia, and asked them to complete various questionnaires about their beliefs, attitudes and emotional states. Of particular interest was the 'fear of compassion' scale, which examines the concerns that we described at the start of this section by asking participants to rate statements such as 'I worry that if I am compassionate, vulnerable people can be drawn to me and drain my emotional resources' and 'Being too compassionate makes people soft and easy to take advantage of'. The people who scored highly on this scale tended to have worse mental health outcomes during the crisis; shutting down their compassion was the opposite of self-protection.[26] We should not be scared to give or receive support in life's darkest periods.

If you currently struggle to act compassionately to people in distress, do not worry: this barrier can be overcome. Like the physical muscles of the heart, compassion for others can grow with intent and practice. The more we use our current reserves today, the more we will have in the future – and the happier we will be.[27]

Based on all these findings, our tenth law of connection can be summarised as follows: **Offer emotional support to those in need, but do not force it upon them. Validate their feelings while providing an alternative perspective on their problems.**

Having learned about the science of solace, it is easy to understand why so many people have continued to find comfort in Henry James's letter to Grace Norton. He acknowledges the strength of her feelings with the sincerest words of sympathy without presuming to know the depths of her pain and shows the greatest humility in his 'ignorance' of her situation. But then he shifts gears and places her suffering in the broader perspective of her life, with an emphasis on the fleeting nature of our feelings, and reaffirms her strength. He concludes by comparing Norton to someone riding an out-of-control horse, and – fearing that she will take drastic action – pleads with her to hold on to her life:

> I insist upon the necessity of a sort of mechanical condensation – so that however fast the horse may run away there will, when he pulls up, be a somewhat agitated but perfectly identical G.N. [Grace Norton] left in the saddle. Try not to be ill – that is all; for in that there is a future. You are marked out for success, and you must not fail. You have my tenderest affection and all my confidence.[28]

Norton did hold on, and the pair remained close confidants until James's death in 1916.[29]

What you need to know

- We underestimate how much people will appreciate our offers of emotional support – and this is especially true for those on the fringes of our social network

- Thanks to the hydraulic model of emotion, we often believe that passive listening is the best way to offer support, as people vent their feelings. Yet research suggests this is often ineffective
- To help people process their feelings, we must validate their experiences *and* help them to reconstrue the situation, so that they can gain more insight into their problems and – where possible – reappraise the negative emotions
- Our caregivers taught us emotional regulation during our childhoods. If we struggle to process difficult feelings, it may be a result of conversations that encouraged catastrophising rather than reappraisal

Action points

- When you hear of someone going through a hard time, reach out and offer your support. The more you practise compassion, the more natural it will become
- Do not try to impose your help on them, but let them know that you are available for help whenever you are needed
- Ask the person what kind of support they would like – if they need to be heard, or if they would instead prefer practical aid, or even a distraction
- When discussing a fraught subject, try to ask *why* questions – which encourage broader insights – rather than *how* questions that may lead people to become caught up in the details of what has happened
- Look for ways of examining the situation from a distanced perspective. You could ask the person what they have learned from the experience or how they view it in the grand scheme of their life. Or you might encourage them to think about how they might view the event from a future perspective. All these techniques encourage more insightful reflection rather than rumination

CHAPTER 11

CONSTRUCTIVE DISAGREEMENT

With today's political polarisation, it is almost impossible to imagine two people who would be less likely to hit it off. She was a liberal feminist who had spent her life fighting for gender equality, and openly supported LGBTQ+ rights. He was a conservative who opposed affirmative action and gay marriage, but consistently favoured gun rights. Yet the judges Ruth Bader Ginsburg (1933–2020) and Antonin Scalia (1936–2016) enjoyed a famously close and long-lasting friendship.

Their bond took root in the 1980s, on the benches of the US Court of Appeals in Washington, DC. They had common ground in their childhoods in New York and shared a love of classical music, and soon developed deep respect for each other's legal ability and work ethic. Within little more than a decade, both had joined the Supreme Court, where their friendship continued to flourish: Scalia was nominated by Republican President Ronald Reagan in 1986 and Ginsburg by Democrat President Bill Clinton in 1993. Outside work, Ginsburg and Scalia frequently shared New Year's Eve celebrations and other holidays; in her chambers, she kept a photo of the two of them on the back of an Indian elephant.[1] They made jokes at each other's expense but were generous with their compliments.[2] They were, Ginsburg said, 'best buddies'.[3]

Their disagreements were fierce, but fair.[4] 'I attack ideas. I don't attack people,' is how Scalia put it.[5] 'Some very good people have some very bad ideas.' If anything, their disagreements only sharpened their legal judgements. 'When I wrote for the Court and received a

Scalia dissent, the opinion ultimately released was notably better than my initial circulation,' Ginsburg said after his passing.[6]

How could two people with such different worldviews form such a strong connection? 'It's hard to remember sometimes that political disagreements, in the not-too-distant past, weren't necessarily cause to retreat into our respective corners, and that ideological differences weren't viewed as moral defects,' the *New York Times* columnist Jennifer Senior has noted.[7]

We could all heed this lesson. Around 40 per cent of registered voters in the US say that they do not have a single friend across the political divide, and each side tends to blame the other for being too closed-minded, unintelligent or dishonest.[8] Confrontations with people of fundamentally opposing viewpoints leave us feeling frustrated, anxious and angry. To preserve your friendships, you could simply avoid discussing contentious topics altogether, but psychological research on self-disclosure tells us that concealment can itself drive a wedge in a relationship. For many people, the only path ahead is to give up on their friendships as soon as a serious disagreement rears its head.[9]

But is there a third way? The psychology of social connection reveals many strategies to enter arguments more constructively, framing disagreements in such a way that we can exchange opinions without shattering a sense of shared reality. At worst, this enables us to better understand each other's positions; at best, we may find that a relationship has been strengthened by the honest and respectful expression of differences, and we may discover that there is more common ground than we once realised.

DIVIDE OR CONFORM?

If you have ever felt that someone else's opinions are so far removed from your own, and yet so confident in their conviction, that one of you must be mad, then you'll identify with Solomon Asch's participants. You may already be familiar with the methods, since his study is now considered a classic of psychological research, though its conclusions are often sorely misinterpreted. The visitors to Asch's lab at Swarthmore College in Pennsylvania were placed into groups of eight and presented with simple tests of perception, such as the following. Which line on the right-hand card – A, B or C – matches the length of the line on the left?

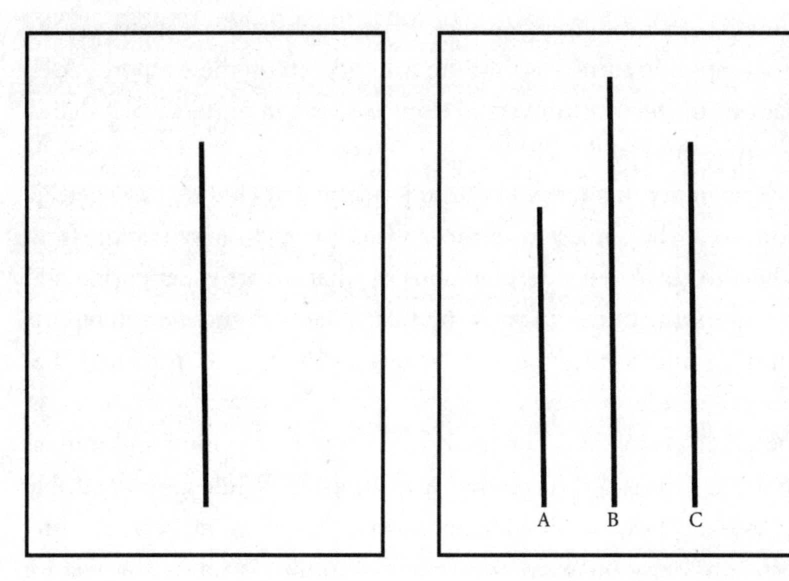

In each group of eight, only one was a real participant, while the rest were actors. For a few trials, the actors all gave the right answer. After that point, however, they started to vote unanimously for the wrong choice – in the example above, either A or B – even though the correct answer was so easily visible. The participant was thus confronted with 'two contradictory and irreconcilable forces: the evidence of his own experience of an utterly clear perceptual fact and the unanimous evidence of a group of equals', Asch later wrote. 'He faced, possibly for the first time in his life, a situation in which a group unanimously contradicted the evidence of his senses.' As you might hope, the participants' trusted their own eyes most of the time, but in around a third of the trials, they ended up endorsing the consensus.[10]

The findings were published in the early 1950s, and over the subsequent decades Asch's study has been considered a textbook example of human conformity as a sign of human weakness.

According to this interpretation, we will happily go along with popular opinion, even if the evidence to the contrary can be found right under our noses – simply because we are too scared or lazy to disagree.[11] For Asch, however, the interest lay in the cost of the disagreements. Questioning the participants after the experiment, he found that almost everyone experienced high levels of anxiety as the group opinion turned against the truth. Based on these reports, Asch was one of the first thinkers to emphasise the importance of a shared reality for our wellbeing.

Remember that many of our judgements – including basic perception – can be subject to error, and so we rely on validation from others to check our interpretations of what we are experiencing and to make sense of the world around us. When another person reports thinking and feeling the same things as us, we feel reassured that our minds are working correctly, and we know that we can accept the other person as a potential partner for cooperation and mutual support. This is threatened when we disagree. While we may be able to explain away small differences in opinion as matters of taste, different views on some fundamental truth – such as the relative lengths of three lines – will lead us to question both ourselves and the other party.

Asch's interviews suggested that the participants who conformed to the majority opinion were not being lazy or weak-willed. The experience had shattered their confidence in their judgement, and they had genuinely questioned whether their perceptions were flawed; they simply didn't know who to trust. Tellingly, Asch found that the presence of one other dissenting voice, who saw things in the same way as the participant, was enough to reduce this stress and their subsequent conformity.[12]

Sadly, Asch's conclusions never really caught on among social psychologists. Following the Second World War, many psychologists were intent on understanding how large populations could be complicit in horrific atrocities such as those committed by the Nazis. In this context, the interpretation of Asch's results that gained the most traction was the simple one: that many people prefer to conform, and that we often act like sheep, mindlessly following the rest of the herd. The more complicated theory – that Asch's participants were highly troubled and struggling to make sense of conflicting information – largely fell by the wayside.[13]

Recently, however, social psychologists have started to revisit the experiment, and they have shown that our existing feelings of loneliness or connection can influence the ways that we respond to disagreement.[14] Consider the work of Elizabeth Pinel at the University of Vermont, who, as we first saw in Chapter 2, has pioneered the scientific study of existential isolation. Like many of the other psychological constructs we have examined in this book, existential isolation can be measured with a questionnaire that asks people to rate various statements on a numerical scale from 0 (strongly disagree) to 9 (strongly agree). People with higher levels of existential isolation are more likely to agree with statements such as 'Other people usually do not understand my experiences', and they are more likely to disagree with statements such as 'I often have the same reactions to things that other people around me do'.

Pinel hypothesised existential isolation should render people more likely to conform to the majority opinion, and in subsequent experiments she showed that this feeling can be manipulated. Simply learning that someone else shared the same thoughts as them on another task helped many participants to resist conformity on the line-perception task, since they already had the reassurance of a shared reality, while those who were deprived of this experience were far more likely to go with the group.

Pinel asked the participants to play the game Imaginiff, for instance. As we saw earlier (see p. 32), this involves questions such as 'Imagine if Michael Jordan were a sea creature. Would he be an octopus, dolphin, hammerhead shark or crab?' People make their choices based on their gut reactions, and hearing that another person has given the same response can create a strong feeling of social connection. Remarkably, this reduces the likelihood of conforming to the group consensus on the line-perception exercise.[15]

These experiments may sound a little far removed from everyday life, but the general conclusions have important implications for our understanding of people's worldviews. They can help us, for instance, to understand why lonely people are often more likely to believe in conspiracy theories.[16] If someone feels isolated from the people around them, they will turn to other potential sources of connection – which may come from an online community of 'independent thinkers' who provide neat explanations for their experience of alienation. To create and maintain the newfound sense of shared reality, they'll be far readier to believe claims that would stretch the credulity of the average person. And misguided attempts to debunk those beliefs may only make things worse.

Let's imagine that your cousin has become friends with some flat Earthers, and that he believes that the presence of globes in education is a worldwide conspiracy. Your immediate reaction may be laughter and ridicule – and why not? He has expressed something that contradicts scientific knowledge and personal experience. 'Have you gone crazy?' you say. 'Any idiot knows that the world is round. We've got photos of the Earth from space! And how else could you explain the fact that we have different time zones? Or the way that ships moving over the horizon disappear bottom first?' You hope that the strong words will shock your cousin into your way of thinking.

The danger, however, is that your mockery will have increased his sense of existential isolation; he'd come to you hoping for understanding and instead you've highlighted just how little respect you have for his view. He may now turn to his new friends to fulfil the need for connection and validation, and because of the increased vulnerability that you provoked, he may be more receptive to even more outlandish theories. Perhaps he'll now entertain the idea that sinister lizard people are secretly running the world. If you really care about your cousin, and want him to listen to reason, your efforts have backfired and you have made it much harder for him to form a more rational worldview.

Now imagine that you respond far more gently. 'That's an unusual idea,' you manage to say despite your disbelief. 'Where did you come across this theory?' Once he's described the YouTube videos, you try to probe deeper. You might ask about these people's academic backgrounds and the reason your cousin believes them to be a credible source. Only then do you try to present your arguments to the contrary, and you do so as a dialogue rather than a diatribe. If you conduct this conversation in the right way, you will have avoided amplifying his existential isolation, setting the stage for a more constructive exchange of opinions. You may not persuade him immediately, but at least you will not have led him to double down on his beliefs.

I have deliberately chosen the extreme example of flat Earth theory to avoid triggering any partisan loyalties, but you may find very similar reactions in the topics of many dinner-table conversations. To take a handful of controversies that continue to occupy the news as I write this chapter: Was Vladimir Putin 'provoked' into invading Ukraine? Is global warming the result of carbon emissions, and does it represent one of the greatest long-term dangers to humanity? And do vaccines weaken rather than strengthen the immune system?

According to a survey by Pew Research undertaken in 2021, 59 per cent of Americans believe that when it comes to important issues facing the public, most people cannot agree on the basic facts. This mistrust in others' knowledge and understanding is even higher in France (61 per cent) and comparable in Italy and Spain (55 per cent each), Belgium (51 per cent) and Sweden (45 per cent). In the UK, the figure is a little lower – 37 per cent – but that still represents a substantial fraction of the population feeling alienated from those of other political backgrounds.[17]

If we want to build authentic and honest relationships with the people around us, we should strive to maintain an open dialogue about the topics that most matter to us. Unfortunately, our conversations frequently descend into bitter disputes that only drive the two parties further apart. As a result, we lose touch with the people who disagree with us, and retreat into friendships with people who will happily reinforce our existing beliefs. Our social network becomes an echo chamber.

We can only avoid this fate if we work to create or maintain a sense of shared reality with the people around us despite the fundamental differences in our core beliefs.[18] And we will find it much easier to do so by following four key principles – **Be civil and curious in disagreements; show interest in the other side's viewpoint; share personal experiences; and translate your opinions into their moral language.** Together, these four principles constitute the eleventh law of connection.

THE MONTAGU PRINCIPLE – CIVILITY WINS ARGUMENTS

Our first principle is named after Lady Mary Wortley Montagu (1689–1782), the English writer and poet who, among other achievements, is best known for having advocated for smallpox vaccinations after her travels in Turkey, and for challenging the prevailing social attitudes towards women. Writing to the Countess of Bute, in May 1756, she advised practising respectful manners to a mutual acquaintance who was delivering her letters: 'Remember civility costs nothing and buys everything,' she implored. 'Your daughters should engrave that maxim on their hearts.'[19]

Linda Skitka at the University of Illinois calls this the Montagu Principle, and her research suggests that it is as true today as it was in the eighteenth century. Displays of bad manners place people on the defensive; sensing an attack on their status, their beliefs and their values, they will look for ways to protect their egos. They will automatically become less open-minded to the information you are trying to impart. Some people may use bad manners as a way of signalling their dominance or the strength of their convictions. But incivility rarely wins respect; more often than not, it reduces others' estimations of your warmth and dominance. You may think that your insults make you look strong, but for most people listening they will only betray your weakness.

By acting uncivilly, you may even lose credit among the people who would be most likely to agree with your stance. In the late 2010s, Skitka and colleagues analysed people's responses to President Donald Trump's tweets. They found that his approval from both his supporters *and* his detractors rose after polite and graceful exchanges, such as 'A fantastic day in DC. Met with President Obama for the first time. Really good meeting, great chemistry. Melania liked Mrs O. a lot!' Public favour fell when he resorted to insults and name-calling, such as 'Crooked Hillary Clinton now blames everybody but herself, refuses to say she was a terrible candidate. Hits Facebook & even Dems & DNC.' Contrary to the theory that attacks on opponents can shore up support from your closest political allies, the researchers found that there was no benefit to the uncivil behaviour, even among Trump's diehard supporters.[20]

While I do not wish to relitigate the 2016 presidential contest, it is worth acknowledging that Clinton also lost popularity among undecided voters when she described Trump's supporters as a 'basket of deplorables'.[21] As far as I am aware, there is no evidence to suggest that one side of the political spectrum demonstrates more civility than the other. Like so many of the laws of connection, the Montagu Principle will apply to people of all backgrounds. Disdain will only poison the discourse and destroy the possibility of establishing mutual understanding.

THE IMPORTANCE OF BEING CURIOUS

Besides showing basic civility, we can make a conscious effort to express genuine interest in what the other person has to say, while acknowledging the doubts and uncertainties in our own views. Following a law of equal action and reaction, signs of humility and curiosity from one side of an argument can help to open minds on the other – with enormous benefits for the ensuing discussion.

Frances Chen at the University of British Columbia has found that asking someone a *single* question to explain their opinions can lead them to view your arguments more favourably and increase their engagement in further discussion. While working at Stanford University, she invited students to engage in an online debate over the introduction of new comprehensive exams before graduation – a move that was likely to provoke lively responses. Unknown to the participants, their debating partners were the experimenters themselves, who followed a careful script arguing in favour of the controversial move. In half of the conversations, this figure asked the students to elaborate on their views. 'I was interested in what you're saying. Can you tell me more about how come you think that?' For the rest of the trials, the conversation continued without any request for more information. And after the debate, the participants were asked to give their views of their partner.

It was a tiny change in the script, but the single expression of interest transformed the students' attitudes to the debate. The participants who heard this question were more willing to receive further information on the opposing arguments and were more agreeable to the prospect of having another conversation with their debating partner on the topic. Analyses of the students' writing on the topic also suggested that they were more willing to take their partners' views on board while formulating their own opinions on the topic.[22]

There are many other ways of signalling our willingness to establish a shared reality across clashing worldviews. Simple expressions of agreement on the points of common ground ('You're right') or acknowledgement ('I understand', 'I see your point') are particularly noted and appreciated during discussions of contentious issues. So are hedging words ('somewhat', 'might') that add nuance to the claims you are making. Finally, there are 'you' statements that attempt to restate what the other person has been saying – demonstrable proof of active listening. These words and phrases may seem inconsequential in the grand scheme of things, but they each help to demonstrate thoughtful consideration of the other's point of view. And they are contagious. When one party starts using more responsive language, the other party is likely to follow suit.

Our off-kilter self-perception means that we're not generally very good at judging how receptive we seem; in many cases, we truly want to engage in a constructive conversation but we're not communicating that intention effectively. Consider a study of local and state government executives who were asked to debate contentious political issues with their colleagues. When judging their own behaviour, the executives focused too narrowly on a handful of signals – whether they swore, whether they used the person's correct title in the conversation, and whether they thanked their partner for their contribution. Yet it was the other linguistic signifiers that affected how they were perceived – things like the deliberate acknowledgement of the other person's points of view, the hedging words that recognise uncertainty and reflect humility, and the focus on points of agreement. Besides influencing the perceptions of receptivity, these small cues influenced how much a person respected their partner's professional judgement, whether they would feel happy with that person representing their organisation, and whether they would consider forming a collaboration with them in the future, all of which would reflect a much healthier and more constructive dynamic for the relationship.

These signs of receptivity can be equally important in online platforms, where they can prevent discussions from escalating into verbal warfare and aggressive pile-ons. Take, for instance, an analysis of Wikipedia editors' behind-the-scenes conversations, which can often descend into bitter arguments. This was far less likely to occur if the posts showed linguistic markers demonstrating respect, humility and genuine interest in alternative points of view.[23]

So why does it work? By emphasising our desire for mutual understanding, receptive dialogue primes a sense of connection, which lowers the other person's self-defences. As a result, they are more likely to listen to our ideas and recognise the biases that might be skewing their thinking.[24] In some cases, this can occur without us even pointing out the flaws in their argument. This surprising shift in perspective could be seen in a group of British students who were asked to consider prejudices towards different groups, such as the elderly or people of colour, before a research assistant opened a discussion on the topic. The research assistant prefaced the conversation with reassurance that any views expressed by the student would be completely confidential: they could be as honest as they wanted without fear of repercussions. With half the students, the research assistant simply listened quietly. They nodded their head and muttered small words of assent (such as 'mmm' or 'I see') but did not actively agree or disagree. For the rest, the research assistant was far more engaged: they asked questions that encouraged the student to open up about their beliefs, and expressed empathy, saying, 'I realise this can be difficult to talk about.'

In neither condition did the research assistant actively challenge the student's views. As if by magic, however, the research assistant's demonstrations of active engagement encouraged the student to think more critically about their prejudice. This resulted in a greater willingness to change their views, and more favourable attitudes to the group in question. To prove that this was a robust phenomenon, the researchers – led by Guy Itzchakov at the University of Haifa – replicated the experiment in a larger sample of Israeli undergraduates. Once again, the research assistant either responded passively to their students, or they showed more active interest, and once again, this shaped the students' capacity for self-reflection. When the research assistant demonstrated curiosity and empathy, the students reported greater insight into the sources of their own bias and a greater willingness to reappraise their views. They were more likely to agree with statements such as 'I feel that I ought to re-evaluate the event now, after the conversation' and 'My conversational partner made me think about the attitude I described during the conversation'.[25]

We can see this principle playing out on the world stage. In surveys examining attitudes to Scottish and Basque independence, a sense of being understood by the other side increased people's feelings of trust and forgiveness towards their opponents. Simply being told that someone has attempted to take your opinions seriously can work wonders, it seems.[26]

Showing uncertainty about your point of view, and demonstrating curiosity about another's, may sometimes feel uncomfortable. If you have very strong opinions about an issue, you may worry that taking a non-judgemental, questioning attitude betrays a lack of integrity; it may even seem that you are simply kowtowing to their views. If your aim is to raise greater awareness and compassion for the causes that you care about, however, the research is clear. Adopting a receptive attitude is not only the best way to maintain a social connection; it's also one of the best methods of changing opinions.

TUCHOLSKY'S PRINCIPLE – SHARE PERSONAL EXPERIENCES

You have probably heard the statement that 'the death of one man is a tragedy, the death of millions is a statistic'. The comment is often attributed to the Soviet leader Joseph Stalin, but an earlier iteration can be found in the writing of the Jewish satirist Kurt Tucholsky, who fled Germany in the 1930s.[27] Much like the Montagu Principle, Tucholsky's assertion is now supported by reams of psychological research, which confirm that anecdotes tend to be much more persuasive than pure data. From a logical perspective, this is rather absurd – we should care far more about a general trend than a single incident – but the concept of shared reality offers one explanation for why that might be. When we scan the brains of people listening to an evocative tale, their neural activity begins to synchronise with the speaker and with other listeners; the words conveying the experience are building the foundations of a mutual understanding.[28] A statistic – without the details of a particular life and all the feelings that person has experienced – is unlikely to do the same.

If someone hears the fact that 46 per cent of LGBTQ+ people have experienced workplace discrimination in the past five years, for instance, that fact may justify stricter legislation to protect their rights. Unfortunately, it can easily pass through the audience's heads without evoking any emotional response or eliciting a sense of connection to the people who are affected. When they hear a real-life story of someone who had been humiliated in their office, however, they will relive the repercussions of those attitudes, relating that person's experiences to their own in a manner that is far harder to dismiss.[29]

Despite the familiarity of Tucholsky's remark, most people do not put this principle into practice. When researchers asked 250 participants to describe the best ways to present their opinion, 56 per cent chose the presentation of facts, while just 21 per cent selected the expression of personal experience.[30] We can see the power of 'Tucholsky's Principle' in a large study of political canvassers working for various liberal organisations in the run-up to the 2018 US midterm elections. Through face-to-face conversations with voters, they aimed to reduce the stigmatisation of illegal immigrants in the US, and they were given one of two potential strategies to do so. Some were told to do so using purely statistical arguments – concerning, for instance, the common fear that immigration increases crime. Others were asked to engage in the exchange of narratives. Given the persuasive benefits of expressing curiosity and interest, these volunteers first asked their conversation partners to describe their own experiences with immigrants and their views on the roles that immigrants play in society. After they had engaged with what the voter had said, the volunteers then offered their own accounts, including descriptions of family members or friends who had entered the US.

Whereas the argument-based conversations had no overall effect on people's views, the inclusion of the narratives managed to shift their opinions. Having exchanged stories with the volunteers, these voters were around 5 per cent more likely to 'strongly support' granting legal status to children who had been brought into the US illegally. That's not bad for an eleven-minute conversation on a contentious topic that most often arouses hostility rather than empathy.

As further proof of principle, the researchers rolled out two further trials – this time examining discussions around trans rights. In total, they recorded more than 6,800 conversations with voters, and the results were always the same: a mutually respectful exchange of experiences was significantly more likely to shift opinion than conversations that focused more on impersonal facts and statistics.[31]

Your political priorities may be very different from those chosen by those canvassers, but Tucholsky's Principle appears to hold for any discussion on any topic; studies have shown, for instance, that exchanging personal opinions reduces knee-jerk responses for people on both sides of the debate around gun control and the right to bear firearms.[32] If just a few more of us take notice of this fact, we may find that our discussions engender far less polarisation and far greater understanding.

MORAL REFRAMING

One final strategy to improve the quality of discussions is moral reframing, which involves identifying your conversation partner's core moral values, and then discussing the issue in question in those terms. Imagine you are exploring environmental concerns with someone who is very patriotic. You might try to list all the terrible things that are happening to the country's ecosystems while also explaining how this is threatening local traditions and customs and many people's livelihoods; preventing pollution and the exploitation of natural resources is the best way to maintain many of the nation's assets and the sanctity of its countryside. By explicitly appealing to this person's love of their country, you may find that they pay far more attention to what you are saying.

To test whether this would work, researchers at New York University and Reed College in Oregon questioned undergraduates about their general political worldviews and then presented them with information about recent ecological research. For half the participants, this contained the following text that was designed to appeal to their patriotism: 'Being pro-environmental allows us to protect and preserve the American way of life. It is patriotic to conserve the country's natural resources.' Afterwards, the participants completed various questionnaires about their environmental behaviour, and were also given the opportunity to sign petitions on the development of 'green jobs', the protection of the US's Arctic National Wildlife Refuge and the prevention of oil spills from offshore drilling. Sure enough, the new framing increased engagement among people with more conservative values, who were traditionally more apathetic about environmental issues.[33]

Moral framing has since been proven to increase open-mindedness for a range of sensitive topics – from same-sex marriage to the choice of the US president. Without changing the facts themselves, you are simply looking into the mind of the other person and explaining the issue in language that they will better understand.[34]

SARAH'S STORY

If you ever fear that some opinions, once formed, will be impossible to change, you might remember the extraordinary story of a woman called Sarah.*

Sarah was a disillusioned American schoolgirl angered by her parents' divorce, alienated from her classmates and confused about her sexuality when she first started mixing with a group of skinheads at high school. Most of them were just 'watered-down punk rockers', she said, but she slowly began to side with a more racist splinter group, and her arms were soon inked with swastikas. 'It was kind of that anger and that violence when I started out that kind of made it very easy for me to fall in with them.'

If she had any seeds of doubt, she crushed them. At one point she embarked on a short-lived affair with a Hispanic man, for instance. She knew she was being a hypocrite – and that she would be considered a 'whore' for sleeping around. But admitting those thoughts 'would have unravelled everything'. On a few occasions, she even packed her belongings to leave town, but a growing sense of existential isolation always drew her back, and each time she tried and failed to leave her commitment to the group grew stronger – an attempt, perhaps, to cling to the brittle sense of shared reality. 'I literally made a point to go out and recruit more people and, you know, to be more hardcore and start more fights.'

---

* To protect this individual's anonymity, her name has been changed.

Sarah's descent into this criminal underworld ended with an armed robbery. The police tracked her down; she was arrested, charged and given a twenty-year prison sentence. Her time in jail could have been another turn in a downward spiral; instead, she experienced something of a miracle.

It began with an act of kindness, when a Black inmate offered Sarah a much-desired cigarette. Sarah didn't change her views immediately – but she was willing to make an exception, and as the relationships developed, she found herself having to defend her opinions to her new friends. 'If you met me on the street before this, would you have kicked my ass?' one woman asked her. 'Those are, you know, the kind of things you can't bullshit your way through an answer,' Sarah reflected. 'As much as they tried to learn about me through the questions they were asking and the answers I was having to give, I was learning about me.' Slowly, her ideology began to crumble, and she ended up testifying against her former gang.

Through her rehabilitation, Sarah eventually met John Horgan, a psychologist at Georgia State University specialising in deradicalisation who documented her journey in a paper for a behavioural sciences journal.[35] After leaving prison, she started working with at-risk youths to prevent others from following her path. 'I started realising the world truly is so much bigger than me and my beliefs and ideas,' she told Horgan. 'It was like being reborn.'

When reading Horgan's account, I'm particularly struck by the non-judgemental nature of Sarah's conversations with her fellow inmates – and how, despite the desperate nature of her circumstances, they embody the principles we have discussed in this chapter. 'They were open and honest and there was no, I want to say there was no tension, but there was no real disagreement there . . . it was just information sharing with not really too much judgment . . . I started becoming the person that I should have been, you know, had I not gone down that path.' Of all the places in the world, prison provided the social environment that she needed to change her mind. 'They knew what I was, and they still treated me like any other person.'

FRIENDS WITHOUT BORDERS

Like the other laws of connection, the four strategies that together comprise our eleventh law must be practised at your own discretion. As I noted at the start of this chapter, clashing worldviews will put enormous strain on any relationship. The research suggests that many people may be more open to reasoned discussion than we imagine, but some will be completely unwilling to engage in a mutually respectful dialogue, no matter how hard you try to remain civil. In such cases, your efforts may be wasted – and I'd suggest that you save your energy for other people. If we are faced with truly toxic views, we certainly shouldn't feel obliged to continue a dialogue that may harm our mental health.

But I hope that these principles may help you to have more constructive conversations with at least a few people in your life who hold different views from your own. As Ginsburg and Scalia showed us, disagreements needn't stand in the way of deep connection; you may even find that rigorous debates add a spark of interest to your friendships and prevent your thinking from becoming lazy and arrogant.

'What's not to like?' Scalia once quipped. 'Except her views on the law, of course.'[36] The two judges may often have come to different interpretations of the Constitution, but they had the same commitment to the truth *as they saw it*, and allowed that to be the foundation of their friendship.

When Scalia sent Ginsburg twenty-four roses for her birthday one year, his friend and colleague Jeffrey Sutton was genuinely perplexed at the gesture. 'What good have all these roses done you?' he asked. 'Name one four–five case of any significance when you got Justice Ginsburg's vote.'

'Some things,' Scalia responded, 'are more important than votes.'[37]

What you need to know

- Feelings of existential isolation can push people to take more extreme views that align them with a particular group
- Profound disagreement over seemingly self-evident truths can exacerbate these feelings of existential isolation. This reinforces the strength of each party's beliefs, rather than changing opinion
- The Montagu Principle – that civility costs nothing and buys everything – is particularly relevant during disagreement. Disrespect reduces your chances of persuading the person in question, and can damage your reputation among the people who would normally support your position
- People have an enormous desire to be heard and understood. When you express interest in another's opinions, you lower their psychological defences, which subsequently opens their mind to your arguments – resulting in a more constructive dialogue
- Personal anecdote is often more persuasive than statistical argument – a phenomenon I call Tucholsky's Principle

Action points

- In any disagreement, avoid swearing, name-calling or any sweeping statements that question someone's personality or capability. In the words of Antonin Scalia, attack ideas, not the person
- Try to validate the other person's sense of self by emphasising the qualities that you most respect in them
- Ask open questions about their points of view and try to reiterate some of the points they have made to show that you have been listening carefully and engaging with their argument
- Recognise any uncertainty or ambiguity in the topic under consideration, and own up to the limits of your own knowledge, rather than conveying a false sense of certainty
- Exchange stories and experiences that can add crucial context to your beliefs

CHAPTER 12

FINDING FORGIVENESS

In 1976, the producer of the comedy TV show *Saturday Night Live*, Lorne Michaels, had a special offer for the Beatles.

'We're being seen by approximately 22 million viewers, but please allow me, if I may, to address myself to four very special people. John, Paul, George and Ringo: the Beatles,' he began. 'In my book, the Beatles are the best thing that ever happened to music. It goes even deeper than that – you're not just a musical group. You're a part of us. We grew up with you. It's for this reason that I'm inviting you to come on our show.' His punchline was the fee. Flashing a cheque, he claimed his channel could pay $3,000 for the reunion.

As it happened, two Beatles were watching the show just a couple of miles away from the *SNL* studio. Paul McCartney and his wife Linda had dropped in to John Lennon's Manhattan apartment that evening when they saw the announcement, and the two men were far more amenable to Michaels' suggestion than he might have imagined. 'We nearly got into a cab, but we were actually too tired,' Lennon later told the journalist David Sheff.[1] The disbanding of the Beatles is one of the most notorious break-ups in music history. How, after all the strife of the past decade, had they arrived at a point when they could happily contemplate a reunion?

In their early days, the Fab Four had been a model of friendship. 'We're all really the same person,' McCartney explained at the height of Beatlemania. 'We're just four parts of the one. We're individual, but we make up one person. We all add something different to the whole.'[2] As the lead songwriters, Lennon and McCartney had an almost uncanny affinity. One was left-handed, the other right, and in rehearsals they would mirror each other's movements like musical doppelgangers. Where other friends might finish each other's sentences, they could complete each other's musical motifs and lyrics, sparking ideas that would result in some of the most memorable songs the world has ever known. There was competition, naturally, but it was self-expanding – ensuring that individually and together their talents could grow and flourish. Unusually for that era, the two men spoke openly about their love for each other.

And then the resentments started to build: disagreements over business arrangements and personal affairs, and conflict over their creative direction and their individual roles within the band. The recriminations continued after the group's official split, in personal conversations, interviews and songs. Yet their connection – however frayed – somehow survived. By the mid-1970s, they had agreed to stop dissing each other, and started meeting in person. The professional reunion did not happen before Lennon's death, but both seriously contemplated the possibility of working together again. 'I was a bit surprised, having heard all the stories of their rocky relationship, how quickly they resumed their warm friendship,' Lennon's girlfriend, May Pang, later recalled. 'Everyone was sweet.'[3]

Lennon and McCartney may have undergone their rift through the lens of the world's media, but the collapse of a social relationship is surely one of the most common human experiences. Between friends, colleagues and lovers, conflict builds and unkind words are said in attack and defence until our bond breaks under the strain. We are left wondering why we formed the connection in the first place.

In the previous chapter we examined how we can deal with political disagreement; now it's time to deal with personal wounds. What can we do to heal a relationship after we have caused offence? And how can we learn to forgive others' transgressions? Some hurt will run so deep that we are better off breaking ties with the guilty party, but mutual acceptance and understanding are possible far more often than our faulty intuitions might lead us to believe.

If most people struggle to form new friendships, it is even harder to maintain them through strife, leading our networks to erode over time. To avoid this fate, we must learn the twelfth and most important law of connection, starting with a new understanding of spite and forgiveness.

ANATOMY OF A GRUDGE

In one of my first jobs, I received a voodoo doll as my Secret Santa, stitched together by a particularly enterprising workmate. Whenever I had been insulted by senior colleagues, he suggested, I should impale it on one of the pins provided, and my frustrations would evaporate. It was a joke, of course, but the cut and thrust of journalism really could be dispiriting. So I'm not ashamed to say that, following some particularly stressful exchanges with certain co-workers, my voodoo doll came to resemble a porcupine – and some of my anger seemed to vanish as a result.

Years afterwards, I discovered that a psychologist called Lindie Liang at Wilfrid Laurier University in Canada had tested the idea scientifically. Participants in Liang's study first had to recall a time when their supervisor had been rude or unfairly negative, or had failed to recognise their hard work. Half of them were then directed to a website depicting a virtual voodoo doll, on which they could take out their frustration with digital pins, pliers or matches; the control group were simply shown a picture of a voodoo doll without the ability to harm it.

In a seemingly unrelated task, the participants were next presented with a series of incomplete words to fill in. Each had two possible solutions – one of which could reveal lingering feelings of resentment. They were asked to complete 'un _ _ ual', for example – which could either form 'unusual' or 'unequal'. Or they saw 'un _ _ st' – which could either spell 'unjust' or 'unrest'. The idea is that people who were ruminating on their mistreatment would be more likely to pick the words associated with injustice. This is a standard test of implicit feelings, and the participants in the control group were indeed more likely than chance to spell out the terms linked to injustice after they had recalled their unfair treatment. This was not the case for those who had stabbed or burned the voodoo doll. These virtual acts of revenge appeared to have calmed participants' sense of being wronged.

It may sound like a silly experiment, but it had a serious purpose. (For this reason, it was awarded an 'Ig Nobel Prize' for 'research that makes people laugh . . . then think'.[4]) Liang wanted to understand why we retaliate when others hurt us, and the use of the virtual voodoo doll offered a safe way to test the effects of revenge in the lab. The experiment underlined how much we value justice in our interactions with others, and how we can redress the balance by lashing out. It is as if our brains are keeping a register of our own trespasses, and those of the people who have trespassed against us, and an act of revenge can help to balance the books, even if our spiteful behaviour is purely symbolic.

There is a good reason that we evolved this way; remaining conscious of others' misdemeanours stops us from behaving like doormats and, in our evolutionary past, would have been essential to maintaining a good position in the group hierarchy. It can also help to restore our sense of agency, leaving us feeling slightly more empowered when we've been made to feel weak. That's why revenge is sweet.

Unfortunately, spite also leaves many other wounds unhealed. When others have insulted us, for instance, we often feel dehumanised. We can see this in the words and phrases that we use: we say that we have been made to 'feel like trash' or 'treated like dirt'. Our sense of humanity can be measured using psychological questionnaires asking people to rate their emotional depth or intellectual sophistication – traits that are meant to define our species compared to other animals. After an upsetting interpersonal conflict, people give themselves much lower scores. It's as if they've lost something vital in their understanding of themselves. And acting spitefully does little to restore it, according to research by Karina Schumann at the University of Pittsburgh and Gregory Walton at Stanford University.

In one experiment, the scientists asked people to recall someone who had hurt them in the past. They were then asked either to write a vengeful letter that aimed to 'get back' at the transgressor and hurt them, expressing all their grievances and anger, or to write a letter of forgiveness that contributed to mutual understanding. They next completed various psychological tests, including a written questionnaire about self-humanity and a modified version of the Inclusion-of-the-Other-in-Self scale that examined their sense of connection to the world in general:

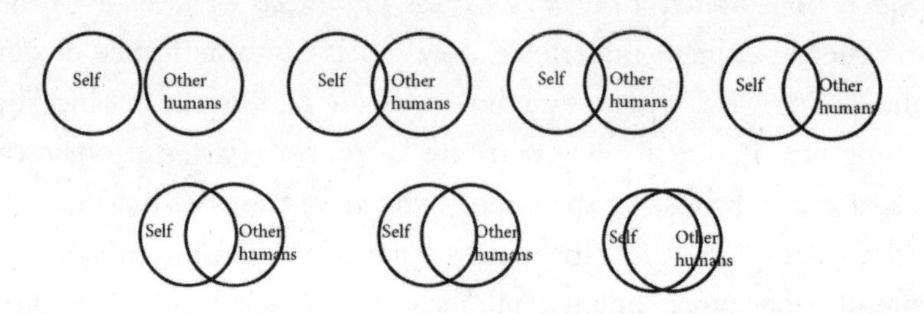

Many of the participants got into the spirit of the experiment. Here's a quote from one person's revenge letter, addressed to someone who had belittled their appearance:

'You said I am ugly? Really? How about you take a look in the god-damn mirror and look at your resented [sic] face? Your nose is big as living hell and your face basically spells ugly in braille with all the acne scars you have on your face. Think twice about crossing me, little one.' Ouch!

And here's a quote from someone trying to forgive their cheating partner:

'Even though I am very hurt by what you have done I realise that mistakes happen. That people make bad decisions and sometimes wish they could take them back. I don't know that this relationship can proceed but I have decided to forgive you and not hold any ill feelings towards you. If this causes our relationship to end I want you to know that I have forgiven you and I hope you can forgive yourself one day.'

I can imagine that the vengeful letters brought a certain satisfaction, but it was the letters of forgiveness that restored the participants' lost sense of humanity and encouraged greater feelings of connection to the world in general. Importantly, their actions were also linked to people's overall view of their own moral character: the people who had practised forgiveness believed that they had lived up to their values, while those who had acted spitefully did not.

The diverse effects of spite and forgiveness could also be seen in people's desire to punish themselves – attitudes that were quantified using yet another virtual voodoo doll. This time, the doll represented not the other person, but *themselves*. When the participants were asked how many pins they would like to stick into its body, the people who had written the revenge letters picked a considerably greater number than the people who had been more forgiving. Previous research on people with depression had shown that responses to this test often reflect real desires to self-harm, so it does appear to be psychologically meaningful. The participants' responses are therefore another sign that we like ourselves a lot more once we have taken the high road, whereas the low road ultimately leaves us feeling dirty and tainted.

Schumann and Walton have replicated these findings in many other experiments. In one study, the participants had to consider a colleague who had criticised a presentation they made at work, for instance, before deciding whether or not to invite that person to a party. In another, they had to imagine either punishing the same co-worker with an unfairly unfavourable peer review, or rating them with scrupulous honesty. In each case, the more forgiving behaviour produced better psychological outcomes.[5]

In addition to restoring our sense of humanity, forgiveness can improve our overall mental health, with studies showing that people who forgive more easily are at lower risk of experiencing depression and anxiety.[6] Like other prosocial acts, forgiveness also comes with physical benefits. Consider a study of 6,671 US citizens in the National Comorbidity Study, who answered the simple question 'Would you say this is true or false? I've held grudges against people for years.' Around 30 per cent answered 'true' – and they were more likely to suffer from cardiovascular disease, chronic pain and stomach ulcers than the people who had answered 'false', even after a host of other confounding factors had been taken into account.[7] Given all we know about the importance of social connection, this makes sense: lingering resentment and conflict is a serious source of stress that would wreak havoc on multiple biological systems, while depriving us of all the benefits of a thriving social network.

The psychological and physical benefits of forgiveness do not negate the importance of boundaries. If someone has mistreated us, we need to make sure that they recognise the consequences of their actions and protect ourselves from further abusive behaviour. Even in the closest of relationships, a proportionate flash of anger can help warn someone that they have overstepped the line and that they need to reconsider their behaviour in the future. Being able to express your frustration or disappointment will also be important for your feelings of emotional authenticity, and – ultimately – for your sense of connection with the other person.[8]

As ever, it's a question of balance. There is a very clear difference between a needless act of spite to wreak revenge on a transgressor, and assertively expressing your frustration or disappointment to ensure that the same insult does not happen again. Aristotle noted as much in his *Nicomachean Ethics*. 'The man who is angry at the right things and with the right people, and, further, as he ought, when he ought, and as long as he ought, is praised.'[9] To be either wrathful or wrathless would result in unhappiness, he argued.

If we hope to have a long-term connection with someone, we need to be open about our feelings, even if they do involve anger and blame. Finding the right ways to express our hurt may be easier said than done – and we'll soon examine some ways to do this constructively. First, however, it may be worth contemplating whether you already have a generally forgiving nature – or if you have a high tendency to hold grudges that may be limiting your life. Think of a time in the past when you have been wronged. On a scale of 1 (strongly disagree) to 5 (strongly agree), how would you rate the following statements?

- I can't stop thinking how I was wronged by this person
- This person's wrongful actions have kept me from enjoying life
- I become depressed when I think of how I was mistreated by this person
- I spend time thinking about ways to get back at the person who wronged me

And:

- I wish good things to happen to the person who wronged me
- I have compassion for the person who wronged me
- If I encountered the person who wronged me, I would feel at peace
- I hope the person who wronged me is treated fairly by others in the future

These items all come from a forgiveness scale developed by scientists studying relationship conflict, and people's answers tend to predict their emotional wellbeing.[10]

When the wound of the offence is still fresh, we may naturally agree more with the first set of statements, representing deep hurt and resentment, than the second set, which reflect greater acceptance and empathy. If, however, we find that aggrieved and vengeful thoughts occupy our minds for long stretches of time, and they represent our attitudes to many different people, then we may have a mental block that is preventing us from forgiving people who might deserve a second chance.[11] This should be of particular concern if we find ourselves adopting these views after a single out-of-character offence. For some, even small resentments can fester, breaking social connections that could otherwise bring great happiness and meaning to their lives. Given the scientific evidence, I don't think it's an exaggeration to say that the lingering bitterness will poison their body and their soul.

People who lean towards this side of the forgiveness scale can learn to let go of their resentments and grudges, with benefits for their mental health.[12] One such technique is built around the acronym REACH, the letters of which stand for *recalling* the pain, *empathising* with the offender, *acting* with altruism, *committing* to your forgiveness and *holding* on to your forgiveness. The third point is essential to this model; rather than seeing forgiveness as something that emerges spontaneously, we can view it as an altruistic gift that we can *choose* to give to someone else.

Through a series of exercises, participants think of a time when they were wounded by another person, and describe how the event affected them. Then they recall a time when they hurt someone else, an act that might help them to have more empathy and compassion for the offender, and a time when they benefited from someone else's forgiveness. One strategy to reinforce this feeling is to write a letter of gratitude to this person. They are then asked whether they would like to grant the same compassion to their offender, and if so, to make the conscious decision to provide the same relief to the person in question. Repeating these steps, when necessary, can increase their commitment to letting go of their resentment.

Everett Worthington at Virginia Commonwealth University pioneered the REACH intervention, which can be delivered through group therapy or DIY booklets, and he tested the approach in multiple studies, including a randomly controlled trial of nearly 5,000 participants from the US and Colombia to South Africa and Ukraine. Half were given the intervention immediately, but the rest stayed on a waiting list until the trial was over (after which they would be free to try the intervention for themselves, if they so wished). The intervention not only improved people's scores on the forgiveness scale, but also soothed symptoms of depression and anxiety.[13]

Worthington practises what he preaches. On New Year's Day in 1996, he discovered that his mother had been murdered by an adolescent who had broken into her home. His initial rage was so intense that he contemplated killing the teen, but eventually he came to forgive the killer.[14] We may struggle to summon the same magnanimity in such circumstances, but Worthington's research suggests we can all learn to be at least a little more understanding of the people around us. If you would like to try the REACH approach for yourself, you can find a link to the DIY workbook in the Further Reading section (p. 249). Forgiveness is never easy, but we need to let go of our grievances, which will otherwise remain lasting barriers to love, friendship and happiness.

ZOOMING OUT

How should we deal with the arguments themselves? Even the most forgiving of us will struggle to keep a cool head in the heat of the moment, only to regret what we have said once the red mist has passed. But psychological techniques can help us to maintain our composure, so that we can express our point of view more constructively, with less danger of permanently breaking the bond.

One strategy revolves around the depth of our focus. When we are angry and hurt, we often concentrate with microscopic attention on the tiny details of events. Under this gaze, the whole landscape of the relationship changes, so that another person's errors come to represent such towering failures that we begin to lose sight of all the reasons that we had forged a connection in the first place. Even the smallest ruptures in our connection can start to take on monumental importance that appears to undermine our understanding of the other person, and the words that we use come to reflect this distorted perception.

After weeks, months or years have passed, we may find that we can zoom out of the situation and put our conflicts into proper perspective; that's why time is such a potent healer. By then, however, it may be too late to salvage our connection and closeness, but certain reappraisal techniques can accelerate this process by helping us to increase the psychological distance from the event at hand. This, in turn, restores a sense of proportion to the errors that have been committed, which makes it far easier to have a constructive conversation.

One of the most striking examples comes from a study of 120 couples from Chicago and its suburbs, who reported their relationship satisfaction, love, intimacy, trust, passion and commitment to each other over two years. For the first year, the couples were left to their own devices, and most experienced a significant decline in the quality of their interactions over this period – a sad but predictable arc for many a promising love story.

At this point, the researchers intervened by giving half the group a very short course on reappraisal. The instructions were simple: look at the conflict through another's eyes. 'Think about this disagreement with your partner from the perspective of a neutral third party who wants the best for all involved; a person who sees things from a neutral point of view,' they were told. 'How might this person think about the disagreement? How might he or she find the good that could come from it?' Having practised applying this technique to a recent conflict for seven minutes, the participants were then encouraged to plan ways that they might use the same strategy in their everyday lives, whenever a disagreement occurred. If I were applying this with my partner, for example, I might decide to take a five-minute walk around the block to cool down after an argument started, during which I'd try to view the discussion through this outside perspective.

To reinforce the idea, the participants were given the same instructions each time the researchers collected their data, at four-month intervals, and they also received a handful of reminder emails in the intervening periods. It was remarkably successful, eliminating the downward trend in relationship satisfaction while those in the control group continued to experience a steady decline in their relationship quality. By viewing their arguments through a different lens and keeping their disagreements in perspective, they were better able to hold on to their love for each other.[15]

If you find it difficult to put yourself in the shoes of a neutral third party, there are many other ways to achieve psychological distance. You could try to conjure up your own future self, for instance. How might you think and feel about the matter of disagreement and your behaviour during the conflict in six months' time, or a year from now? Inspired by the earlier study, psychologists at the University of Waterloo in Canada asked their participants to use this form of reappraisal when thinking about a recent disagreement with a friend or romantic partner. As hypothesised, the small mental shift resulted in greater forgiveness and less blame towards the person who had hurt them. And this, in turn, helped them to see the growth that had emerged from the argument, resulting in greater optimism about their later wellbeing.

'The experience really helped us bond,' one participant wrote. 'It was really troublesome at the moment, but I will definitely see it as a deeper moment between us . . . allowing us to get to know each other and to learn how we both deal with hard situations.' The idea that conflict can bring us closer is hard to remember in the heat of an argument – but adopting a future perspective may help to bring wisdom when we most need it.[16]

As we contemplate a conflict, we may also choose to reflect on our core values and personality characteristics that make us proud. The aim here is to bolster our 'self-integrity', which psychologists define as the concept of ourselves as moral and worthwhile people. When our self-integrity is threatened, we may feel that our very identity is crumbling, and so we attempt to protect it through any means possible. We may lash out at the people who are criticising, or simply shut down and avoid talking to them. Neither is a recipe for reconciliation.

By strengthening our self-integrity, we can neutralise these defensive behaviours, so that we are better prepared to make the concessions that are necessary to repair the relationship. This may sound rather abstract and theoretical, but many studies have shown that it works remarkably well during real-life conflicts. Here's how to put it into practice.

Rank the following eleven values in order of their importance to your identity:

- Independence
- Sense of humour
- Social status
- Learning and gaining knowledge
- Relations with friends/family
- Spontaneity/living life in the moment
- Athletic ability and fitness
- Work ethic
- Musical ability and appreciation
- Physical attractiveness
- Romance

Now think about your number one choice and take a few minutes to explain why it is so important to you.

Even if you do not have the time to complete the full exercise every time you have a conflict, you can try to take a few moments to reflect on the general principles of the strategy. Ask yourself: what are the important values that guide your life? And how do these relate to the conflict at hand? When you look at the bigger picture of your life and the things that you care about, you may find that the disagreement shrinks in importance – and the other person's criticisms are no longer as threatening to your self-integrity. When Karina Schumann at the University of Pittsburgh gave couples these instructions, they managed to have more constructive conversations about recent disagreements. The benefits for their overall relationship satisfaction could be seen a full year after the initial intervention.[17]

If your partner is willing, you may find that a change of scenery helps to put you both in a better frame of mind – perhaps by taking a walk together in a park or garden. Walking is a joint activity that naturally generates a shared focus of attention on your surroundings, which can help build the basic sense of shared reality that was so fundamental at the start of your relationship. Remember that at the most basic neurological level, synchronous movements can help to align our brainwaves. The ensuing sense of connection may just ease your conversation, so that you are both more open to each other's points of view.[18]

Whichever strategy you choose to use, you cannot expect resolution to arrive immediately. Instead, these tools should simply allow us to survey the terrain of the disagreement with greater objectivity, without being blinded by our feelings. This additional clarity may help us to make allowances for other people's mistakes, while articulating our concerns more compassionately. And when we are undoubtedly in the wrong, they will put us in the best frame of mind to heal the hurt that we have caused.

WHY SORRY SEEMS TO BE THE HARDEST WORD

We started with an anatomy of a grudge, so let's finish by investigating the anatomy of an apology. After we have committed a transgression, we need to repair our shared reality with the person in question – and the only way of doing this is to take responsibility for our actions and demonstrate that we fully understand the reasons for their hurt feelings. By avoiding those conversations, we are sabotaging our own happiness and that of the other person.

Psychological research suggests that there are four main obstacles to apologising effectively: we don't appreciate the harm we've caused; we assume that the act of apologising itself will be too painful and shaming; we believe that the apology will do little to repair the relationship; and finally, we may simply not understand what constitutes a good apology – so we fail to say the words that will be necessary for healing.[19] The first point clearly depends on the details of the offence itself, though the psychological distancing and self-affirmation strategies we explored in the last section may help us to overcome this. But the next two concerns – like so many of our assumptions about relationships – are largely unfounded, and they will therefore present unnecessary barriers to social connection.

Researchers from the universities of Cambridge and Rotterdam have shown that we severely overestimate how stressful and humiliating our apologies are going to be, while underestimating how liberating it can be to make amends for our transgressions. For one experiment, the scientists tricked participants into believing that they had cheated on a general knowledge quiz, a behaviour that deprived another participant of the prize. After the event, some of these participants were asked to predict how they would feel if they apologised for their actions, while others were given the chance to say sorry in real life. The apology was not nearly as stressful or humiliating as the participants had predicted. To see whether the same patterns played out in daily life, the researchers asked one group of participants to recall a time when they had apologised to another person – while others imagined doing so in the future. Once again, the predictions were off. Giving an apology was much more pleasant, and less awkward, than people imagined it would be.[20]

There are, in fact, many individual benefits of apologising, besides gaining the forgiveness of the other person. Offering an apology boosts people's sense of their integrity, for instance. It also promotes greater self-compassion, and allows the transgressor to feel more positive about their future behaviour – both of which are important for long-term wellbeing.[21]

The third concern – that apologies will do little to repair *others'* opinions – is also misguided. When they have been hurt, most people are desperate for a recognition of the wrongdoing, and the absence of an apology can sometimes be more upsetting than the initial offence. But we underestimate this desire, and instead believe that our pleas will fall on deaf ears. It is for this reason that people are more likely to make amends when they already have reason to believe that the other person is ready to forgive: they have been reassured that their apology will not be in vain. If we were only a little more willing to put ourselves on the line, we might save many more of our relationships.[22]

This brings us to the content of the apology. It should go without saying that we have to take responsibility for the offence without shifting the blame onto someone else; there's nothing more annoying than 'I'm sorry if you took my words the wrong way' or the passive

'Mistakes were made'. These words only highlight the lack of understanding between us, creating further damage to the already fractured sense of shared reality.

According to multiple social psychologists, the ideal apology covers the following elements:

- A clear statement framing the apology
- Acknowledgment of responsibility for the offence
- Identification of the offence
- A request for forgiveness
- An expression of regret or sadness
- An offer of repair
- An explanation for why the offence occurred
- A promise that it will not recur

The benefits are cumulative: simply stating an apology has some power, but the more elements that you include, the more your sincerity will shine through, and the greater the chance that the offended person will find peace and forgive you.[23]

The timing of an apology can be important. Extended delays will prolong someone's suffering, but we don't want to be too hasty to say sorry, either; otherwise, it can seem like we haven't reflected on our behaviour and made sufficient effort to see the situation from their point of view. At worst, it may seem like we are responding as a reflex, or attempting to rush through the process to avoid hearing the consequences of our actions: maybe we want the reconciliation without the pain of listening to their account. For these reasons, we should allow the other person to fully express their grievances, before we go into the details of our apology. As the authors of one study put it, apologies are a surprising case of 'better late than early'.[24]

In the ensuing conversation, we may also attempt to put the other person's feelings into our own words. Some caution will be necessary. As we have seen previously, our egocentric biases may lead us to overestimate how much we understand the other person's emotions, so we must be careful not to presume too much insight. Provided that we have paid careful attention to what they have said, however, reiterating their points will show that we have accepted the hurt that we have caused and that we wish to take away that pain. The result is a greater feeling of trust, which is essential for forgiveness.[25]

Whenever and wherever we choose to deliver them, our apologies need to signal the relationship's value to us. Much of the hurt of the offence will have come from the feeling that we are no longer invested in maintaining the bond, and we need to put those doubts to rest.[26] Depending on the nature of the offence and the relationship, this should be a time to reiterate our respect and/or love for the other person and why we wish for them to remain close. Only then can the wounds begin to heal as we re-establish our old bond.

ALL WE NEED IS LOVE

Our twelfth law of connection reads as follows: **For your wellbeing, choose forgiveness over spite. Look at the big picture in arguments. Ensure your apologies define the offence, take responsibility for your actions, and express regret.** Since every act of forgiveness rests on an underlying belief – that people can improve – we must add one final clause: **Have faith that people can change for the better.**

Some people assume that our behavioural patterns are fixed in childhood and will never change. For those who hold this view of human nature, a mistake is a sign of serious character flaws that can never be undone. With no chance of redemption, any challenge to their self-integrity will feel even more threatening. And others' apologies will sound hollow and empty, since they are doomed to make the same errors again and again. This cynical view is at odds with the latest neuroscience and psychology, which show that people really can change their ways, if they have the will to do so. The bad habits of someone's past do not have to predict their future. By denying this fact, we prevent ourselves from growing with the people we love, and we will become much lonelier as a result.

Perhaps we should take a leaf out of Lennon and McCartney's (song)book. They may not have had the opportunity to re-form the Beatles, but they did learn to overcome their differences and appreciate each other's company again. When they met up in New York in the 1970s, onlookers were often surprised by how quickly they resumed their old rapport. As their biographer Richard White wrote, 'The warmth that existed between the two men was clearly deeper than any petty residual vendettas.'[27]

By the mid-1970s, Lennon had come to see that much of the bitterness had arisen from his fear of taking his own path without his friends at his side. 'However much we wanted to be independent, it's quite hard to be independent after ten years of being locked in each other's arms, as it were,' he said. 'But I think we've got over that. We're going to be friends for the rest of our lives.'[28] To this day, McCartney argues that the rift had been overblown by the media. 'The story about the break-up, it's true but it's not the main bit,' he told a British chat show in 2014. 'The main bit was the affection.'[29]

What you need to know

- When others treat us unfairly, we lose our sense of agency and our recognition of our own humanity
- After a transgression, revenge helps to restore a sense of justice but it leaves us feeling dehumanised and disconnected from our moral values
- Forgiveness is tightly linked to better mental and physical health – and we can all learn to cultivate it in our lives
- Acknowledging our errors can threaten our self-integrity – the concept that we are a moral and effective person. To avoid this discomfort, we become self-defensive, which reduces our chances of reconciliation
- Increasing our 'psychological distance' from the conflict can help us to

gain a sense of proportion and perspective, allowing us to be more constructive in our conversations

- People overestimate the discomfort of an apology, and underestimate how much better they will feel for making amends. We also underestimate how much our words will mean to the other person. Both these biases prevent us from rebuilding connection after a conflict

Action points

- If you hold on to grudges and struggle to let go of resentments, consider taking the REACH forgiveness programme. It may strengthen your social connections and improve your mental health
- To build psychological distance from the event, imagine what a neutral third party would say about the disagreement. Or consider how you'll view your behaviour in six months' or a year's time
- When apologising, ensure that you give the other person plenty of opportunity to express their hurt and sorrow. The first step to forgiveness is feeling that you've been heard, and if your apology is hasty and rushed, you'll deprive the person of this opportunity
- Be specific about the offence you have committed, accept responsibility, and express sorrow and regret for the hurt that you have caused. Where possible, offer to make amends for any damage you have done, and politely request the gift of forgiveness

CONCLUSION

THE THIRTEENTH LAW OF CONNECTION

We began this journey with the meeting of Helen Keller and Anne Sullivan. Decades later, their connection was as strong as ever. In March 1917, Sullivan was having a five-month break in Puerto Rico, following a painful separation from her husband. Keller missed her greatly during this period, and she must have felt the separation more keenly than ever on the anniversary of their meeting. On that day, she said, she had been 'born again'; she considered it to be her 'soul-birthday'.

'Just think, last Friday was my soul-birthday, and I had to spend that day of days away from you!' she wrote to Sullivan on 4 March that year. 'Thirty years ago you came to a quiet village, you, a young girl all alone in the world, handicapped by imperfect vision and want of experience – you came and opened life's shut portals and let in joy, hope, knowledge and friendship.'[1]

Keller was thirty-seven at the time she wrote the letter. In our society, we mark wedding anniversaries as if marriages are the only meaningful relationships, but having learned so much about the power of social connections, I can't help but wish we made more effort to celebrate our many other ties, remembering those 'soul-birthdays' when others entered and enriched our lives.

In this book, we have learned of twelve laws of connection that help us to establish and maintain stronger relationships:

1. Be consistent in your treatment of others. Avoid becoming a stressful frenemy.

2. Create a mutual understanding with the people you meet. Ignore superficial similarities and instead focus on your internal world, and the peculiar ways that your thoughts and feelings coincide.

3. Trust that others, on average, will like you as much as you like them, and be prepared to practise your social skills to build your social confidence.

4. Check your assumptions; engage in 'perspective-getting' rather than 'perspective-taking' to avoid egocentric thinking and misunderstandings.

5. In conversation, demonstrate active attention, engage in self-disclosure, and avoid the novelty penalty, to build mutual understanding and contribute to the merging of our minds.

6. Praise people generously, but be highly specific in your words of appreciation.

7. Be open about your vulnerabilities, and value honesty over kindness (but practise both, if possible).

8. Do not fear envy. Disclose your successes but be accurate in your statements and avoid comparing yourself to others. Enjoy 'confelicity'.

9. Ask for help when you need it, in the expectation that your pleas for support can build a stronger long-term bond.

10. Offer emotional support to those in need, but do not force it upon them. Validate their feelings while providing an alternative perspective on their problems.

11. Be civil and curious in disagreements; show interest in the other side's viewpoint; share personal experiences; and translate your opinions into their moral language.

12. For your wellbeing, choose forgiveness over spite. Look at the big picture in arguments. Ensure your apologies define the offence, take responsibility for your actions, and express regret. Have faith that people can change for the better.

To these, I would like to add one final law:

13. Reach out to the people who are missing in your life. Let them know that they are still a part of your thoughts.

In this age of globalisation, most of us have friends, colleagues and relatives who cannot be physically present in our lives. As Keller and Sullivan showed, physical distance does not need to equal social distance. We now have incredible technologies that can allow instant communication; we do not need to wait for a letter to cross the ocean to tell someone that we are thinking of them.

As with all the other laws of connection, robust scientific evidence supports this advice, thanks to a series of studies by Peggy Liu at the University of Pittsburgh. Liu's team recruited dozens of students and asked them to name a social tie with whom they'd started to lose touch. In a procedure that will now be familiar to you, the researchers asked them first to write a short note reaching out to this person, and then to complete a psychological questionnaire that measured how much they thought the other person would appreciate the gesture. The researchers then emailed the note to the named person, along with a questionnaire examining their feelings about the message.

Consistent with the research on all our other social biases, the participants significantly underestimated how warmly the other people would respond to their messages – which were almost universally welcomed. Indeed, when asked to rate their appreciation from 1 (not at all) to 7 (a great extent), the average response from the receivers was 6.2. Subsequent research showed that the gesture was especially welcomed by weak ties – those looser connections who expected to have been forgotten and were pleasantly surprised to have been remembered and cherished.[2] Lost friendships are among our most common regrets – and many of us are too proud to make the first move – but even the smallest gestures will help to revive a weakened or fractured bond.[3]

THE FUTURE OF FRIENDSHIP

While discussing this book with my acquaintances, I am often asked whether the medium of contact matters. Do we have to meet face to face to get the full benefits of friendship? And is technology increasing the distance between us? My friends point to press stories claiming that social media is driving people to isolation and depression, as we sit alone with our screens instead of meeting in the real world. 'We're in a loneliness crisis: Another reason to get off your smartphone' as the *New York Times* put it in 2022,[4] the same year Al Jazeera ran a story headlined 'A toxic feed: social media and teen mental health'.[5] The year before, the *Washington Post* ran an article called 'Teens around the world are lonelier than a decade ago. The reason may be smartphones'.[6]

With messages such as these, it's little wonder that some of my friends are concerned about their screen time, while also struggling to wean themselves off their smartphones. That's not to mention worries about remote work. How is it possible to bond with teammates when you only visit the office one day a week?

Their concerns are mostly unjustified. While it is true that too much time spent on social media *can* amplify our feelings of loneliness, the effect depends on the way you use the technology – nuances that have been lost in many of the media stories. Some people spend their time online engaging in social comparison: they see popularity online as a kind of competition and may end up feeling inadequate when they see others' posts depicting their desired lifestyle. Such behaviour is unlikely to benefit our wellbeing. Others use social media as another tool for connection: keeping up to date with their contacts' lives, so that they can offer emotional support during bad times and enjoy confelicity in the good.[7]

The same goes for technology like video conferencing: it can be a powerful tool for maintaining connection, but our flawed intuitions may prevent us from making the most of it. One recent study asked participants either to reconnect with an old acquaintance, or to speak to a stranger online. Most feared that voice or video calls would be undesired and awkward – they expected that people would be much more receptive to text messages and emails instead. When the participants were encouraged to use voice and video calls, however, they found that the conversations were much smoother and more enjoyable than they had anticipated – and they resulted in stronger bonds afterwards.[8]

All other things being equal, face-to-face meetings are probably better than voice or video calls, which are slightly better than emails, texts or DMs on social media, but social contact through any medium is better than no contact at all. Activities such as self-disclosure are likely to benefit your friendship however you practise them.[9] One study has even tested the fast friends procedure in virtual reality, and found increased social connection between strangers as a result of the conversations.[10] Provided that you still apply the laws of connection in your interactions, you should find your friendships flourishing online and offline. 'The deepest principle in human nature is the craving to be appreciated,' wrote William James in the 1890s – and that will remain true through the twenty-first century and beyond, through whichever means we choose to communicate.[11]

If you have not yet practised any lessons from this book, try to think of three people who mean something to you and let them know how you feel. You may be surprised by their responses – a first taste, I hope, of the happiness that can come from applying the laws of connection.

FURTHER READING

SELF-COMPASSION

Professor Kristin Neff, at the University of Texas at Austin, has compiled a comprehensive website full of resources on self-compassion and the ways to cultivate it: https://self-compassion.org/

THE FAST FRIENDS PROCEDURE

You can read all thirty-six discussion topics of the fast friends procedure at the University of Berkeley, California's Greater Good Science Center: https://ggia.berkeley.edu/practice/36_questions_for_increasing_closeness

THE NARCISSISM TEST

The Narcissistic Personality Inventory is available for free at: https://openpsychometrics.org/tests/NPI

FORGIVENESS

Professor Everett Worthington at Virginia Commonwealth University designed and tested the REACH forgiveness program. He offers many free workbooks at his personal website: http://www.evworthington-forgiveness.com/reach-forgiveness-of-others

GLOSSARY

Amae

An 'inappropriate' request for help with a task that we may be able to do on our own, but which will help underline our bond with the other person. Despite the inconvenience it can create, *amae* deepens the sense of social connection for both the person asking for, and the person helping with, the request.

Ambivalent relationships

These members of our social network are highly unreliable; they help and hurt in equal measure, and their unpredictable behaviour can trigger stronger stress responses than purely aversive acquaintances. They may be 'frenemies', colleagues, parents, siblings, spouses, or members of your extended family.

Beautiful mess effect

We fear showing our flaws, yet appreciate others' candour about their vulnerabilities. The beautiful mess effect describes how the honest acknowledgement of our failings can bring about greater social connection.

Benjamin Franklin effect

Asking for help from another person often increases their regard for us – a finding inspired by the observations of Benjamin Franklin. See also ***Amae***.

Benign envy

A constructive form of jealousy that motivates us to change our circumstances, without arousing rancour towards the people who are more fortunate than ourselves.

Boomerasking

The habit of posing a question as an excuse to talk about yourself. This is not conducive to social connection.

Closeness-communication bias

We tend to be overconfident in our capacity to take the perspective of our friends or family members, whereas we are much humbler about our knowledge of total strangers. The closeness-communication bias may explain why we often take our loved ones for granted.

Co-brooding/Co-reflection

Co-brooding is the act of mutual venting, which can exacerbate our unhappiness. When discussing difficulties with others, we should instead aim for co-reflection, which involves looking for new insights and perspectives on the problems at hand.

Collective effervescence

The sense of connection and meaning that comes from group events, when large numbers of people feel and act as one.

Confelicity

A shared joy at another's success or wellbeing.

Egocentric thinking

Our tendency to assume that other people are experiencing the world through our perspective. See also **Illusion of transparency, Illusion of understanding.**

Emotion coaching

A supportive dialogue that helps another person to reappraise their pain or distress.

Existential isolation

The sense that no one else shares our experience or comes close to understanding it. We can experience existential isolation even when we are in the company of many other people.

False consensus effect

A form of egocentric thinking that leads us to think our beliefs are far more commonly shared than they actually are. It can be a source of serious embarrassment at dinner parties.

Fast-friends procedure

A scientific technique designed to engineer intimacy between two people, the fast-friends procedure uses a series of thirty-six questions that encourage increasing self-disclosure.

Gratitude gap

We consistently underestimate how much others will value words of thanks and appreciation, while loving praise ourselves.

Hazlitt's law

To build social connection, we must pay active and visible attention to the people around us. Named after the nineteenth-century essayist William Hazlitt, who argued that 'the art of conversation is the art of hearing as well as of being heard.'

Hubris hypothesis

Boasts can be easily forgiven, provided that they do not make explicit comparisons to others.

I-sharing

An intimate form of shared reality in which we experience the same reactions as another person to a particular event, as if we are occupying the same stream of consciousness.

Illusion of transparency

Our tendency to overestimate how visible our emotions are to the people around us.

Illusion of understanding

Our tendency to overestimate how well others can understand the meaning behind our ambiguous messages, and how well we can understand theirs.

Liking gap

A common bias in which we assume that we like another person a lot more than they like us. The liking gap can discourage us from building on an initial spark or sense of connection.

Meta-perceptions

Our perceptions of how other people perceive us. They are often needlessly pessimistic.

Mitfreude

The opposite of *Schadenfreude*, this translates as 'joying-with'. See also **Confelicity**.

Montagu Principle

Maintaining civility during heated political arguments promotes greater support from both those who agree and those who disagree with your point of view. Named after Lady Mary Wortley Montagu, who first asserted that 'civility costs nothing and buys everything'.

Moral reframing

The act of describing issues in the language of your opponents' moral values during political disputes. Moral reframing is a powerful tool of persuasion.

Novelty penalty

People often find it hard to process new information about unfamiliar ideas or experiences, and so they prefer to talk about things that they already know. This is the novelty penalty and it is a common barrier to social connection in conversation.

Paradox of compassion

Having high levels of empathy for another's pain, and supporting them in their hour of need, can lead to far greater happiness than shielding ourselves away from the suffering of other human beings.

Perspective-getting

Many of us believe that we can gauge others' thoughts, feelings or beliefs by simply imagining ourselves in their shoes – the process of *perspective-taking*. But our assumptions are often embarrassingly inaccurate. For this reason, we should try harder to practise *perspective-getting*, by actively asking people for their opinions.

Perspective-taking

See **Perspective-getting**

Porcupine's dilemma

Schopenhauer's assertion that the closer our relationship to someone, the greater the pain they will inflict on us. We can overcome the porcupine's dilemma by applying the laws of connection.

Self-disclosure

The act of revealing personal or private information about one's self to other people, including our fears, fantasies, and aspirations. Offering greater self-disclosure, and encouraging it in return, can put us on a fast track to closer relationships.

Self-expansion
The feeling of individual growth that is essential for the maintenance of healthy and supportive relationships.

Shared reality
To feel a shared reality with someone, you have the strong sense of experiencing the same inner life – so that your thoughts, feelings and beliefs coincide. It is the foundation of social connection. See also **I-sharing, Existential isolation**.

Tucholsky's Principle
When debating political issues, describing personal experiences is considerably more persuasive than sharing facts and statistics. Named after the German satirist Kurt Tucholsky.

Witnessing effect
Simply observing gratitude directed at another person can promote altruism in onlookers, and primes warmer feelings towards the people giving and receiving the praise. The witnessing effect is an easy strategy to attract a more supportive social network.

ACKNOWLEDGEMENTS

Every book is a collective endeavour, and *The Laws of Connection* could not exist without the kindness and support of countless people. Thank you to my agent Carrie Plitt, for seeing the potential in this idea and for her constant guidance. I'm also indebted to Michele Topham and the rest of the team at Felicity Bryan Associates, and to Zoë Pagnamenta in New York for ensuring that the book reaches readers on both sides of Atlantic.

I'm enormously grateful to the wonderful team at Canongate: Simon Thorogood, Lucy Zhou, Caitriona Horne, Jenny Fry, Alice Shortland, Leila Cruickshank and Claire Reiderman. I'm proud to be published by you. In the US, I've been so lucky to find a home at Pegasus Books, with the support of Claiborne Hancock, Jessica Case and Julia Romero. Thanks also to Fraser Crichton, my copyeditor, and Lorraine McCann, my proofreader, for polishing the text.

I am indebted to the scientists who have shared their expertise on the psychology of conversation and social connection with me over the years, including Vanessa Bohns, Erica Boothby, Taya Cohen, Gus Cooney, Kevin Corti, Nicholas Epley, Alex Gillespie, Naomi Grant, Igor Grossmann, Julianne Holt-Lunstad, John Horgan, Karen Huang, Boaz Keysar, Ethan Kross, Elizabeth Pinel, Lindie Liang, Elaine Reese, Gillian Sandstrom, Karina Schumann, Ashley Whillans and Wouter Wolf.

My idea for *The Laws of Connection* emerged from pieces commissioned by Tiffany O'Callaghan at *New Scientist*, and Meredith Turits at the BBC. Thank you both for sowing the seeds that eventually grew into this book.

Thanks to my friends, colleagues and editors for supporting me and my work, including Sally Adee, Lindsay Baker, Shaoni

Bhattacharya, Jules Brown, Amy Charles, Dan Cossins, Eileen and Peter Davies, Catherine de Lange, Kate Douglas, Stephen Dowling, Natasha and Sam Fenwick, Richard Fisher, Alison Flood, Alessia Franco, Rob Freeman, Alison George, Zaria Gorvett, Richard Gray, Claudia Hammond, Jessica Hamzelou, Sophie Hardach, Martha Henriques, Melissa Hogenboom, Christian Jarrett, Rebecca Laurence, Fiona Macdonald, Ian MacRae, Damiano Mirigliano, Will Park, Ellie Parsons, Emma and Sam Partington, Jo Perry, Dhruti Shah, David Shariatmadari, Mithu Storoni, Neil and Lauren Sullivan, Jon Sutton, Helen Thomson, Ian Tucker, Gaia Vince, Richard Webb and Clare Wilson.

I owe more than I can describe to my parents, Margaret and Albert. Thanks most of all to Robert Davies. It is impossible to imagine my life without you.

CREDITS

Quotation from *The Journals of Sylvia Plath* reproduced courtesy of Faber & Faber Ltd. Quotations from Helen Keller reproduced courtesy of the American Foundation for the Blind.

p. 15 Effects of social connection on health. Based on Holt-Lunstad, J., Smith, T.B., & Layton, J.B. (2010). Social relationships and mortality risk: a meta-analytic review. *PLoS Medicine,* 7(7), e1000316.

p. 33 Inkblot. By kind permission of Elizabeth Pinel. From Pinel, E.C., Long, A.E., & Huneke, M. (2015). In the blink of an I: On delayed but identical subjective reactions and their effect on self-interested behavior. *The Journal of Social Psychology, 155*(6), 605–16.

p. 37 Inclusion-of-Other-in-Self test. Based on Aron, A., Aron, E.N., & Smollan, D. (1992). Inclusion of other in the self scale and the structure of interpersonal closeness. *Journal of Personality and Social Psychology,* 63(4), 596–612.

p. 72 Sally–Anne test. Author's own.

p. 73 The director test. Based on Savitsky, K., Keysar, B., Epley, N., Carter, T., & Swanson, A. (2011). The closeness-communication bias: Increased egocentrism among friends versus strangers. *Journal of Experimental Social Psychology,* 47(1), 269–273. Illustration copyright (c) Tom Holmes, tom-holmes.co.uk

p. 228 Modified version of Inclusion-of-Other-in-Self test. Based on Schumann, K., & Walton, G.M. (2022). Rehumanizing the self after victimization: the roles of forgiveness versus revenge. *Journal of Personality and Social Psychology, 122*(3), 469.

NOTES

Introduction

1 Keller, H. (1903). *Optimism: An Essay*. T. P. Crowell. Available to access online at: https://www.disabilitymuseum.org/
2 Keller, H. (2017). *Story of My Life* (pp. 12–16). Grapevine. Kindle Edition. Originally published in 1903.
3 Quoted in Herrmann, D. (1998). *Helen Keller: A Life* (p. vii). Knopf.
4 Keller, H. *Story of My Life* (p. 75).
5 Holt-Lunstad, J. (2021). Loneliness and social isolation as risk factors: The power of social connection in prevention. *American Journal of Lifestyle Medicine*, *15*(5), 567–73.
6 See, for example: Perry-Smith, J.E. (2006). Social yet creative: The role of social relationships in facilitating individual creativity. *Academy of Management Journal*, 49(1), 85–101.
7 This is the UCLA Loneliness Scale: Russell, D., Peplau, L.A., & Ferguson, M.L. (1978). Developing a measure of loneliness. *Journal of Personality Assessment*, 42, 290–4.
For a summary of these results over recent years: https://newsroom.thecignagroup.com/loneliness-epidemic-persists-post-pandemic-look; Office of the Surgeon General. (2023). Our Epidemic of Loneliness and Isolation: The US Surgeon General's Advisory on the Healing Effects of Social Connection and Community. Available at: https://www.hhs.gov/sites/default/files/surgeon-general-social-connection-advisory.pdf
8 Vincent, D. (2020). *A History of Solitude* (p. 222). John Wiley & Sons.
9 Keller, H. *Story of My Life* (p. 20).

CHAPTER 1: *The Social Cure*

1 After Ghinsberg's rescue, he discovered that Ruprechter was a criminal wanted by Interpol who had previously placed other explorers at risk.
2 The details in this section are drawn from Ghinsberg, Y. (1993). *Back to Tuichi*. Random House; Round, S. (2012). 'I was lost in the Amazon

jungle'. *The Jewish Chronicle*: https://www.thejc.com/news/all/i-was-lost-in-the-amazon-jungle-1.3956

3 McCain, J. (1999). *Faith of My Fathers* (pp. 206–11). New York: Random House.

4 Lewis, C. S. (1960). *The Four Loves* (p. 103). Harcourt, Brace.

5 Details of the Alameda County study can be found in Berkman, L.F. and Breslow, L. (1983). *Health and Ways of Living: The Alameda County Study*. Oxford University Press, New York; Schoenborn, C.A. (1986). Health habits of US adults, 1985: the 'Alameda 7' revisited. *Public Health Reports, 101*(6), 571; Stafford, N. (2012). Lester Breslow. *British Medical Journal, 344*, e4226.

6 Berkman, L.F. (1979). Social networks, host resistance, and mortality: a nine-year follow-up study of Alameda County residents. *American Journal of Epidemiology*, *109*, 189–201.

7 Berkman and Breslow. *Health and Ways of Living: The Alameda County Study* (pp. 200-3).

8 For excellent overviews of these findings and a discussion of their role in public health guidance, see: Holt-Lunstad, J. (2021). The major health implications of social connection. *Current Directions in Psychological Science*, *30*(3), 251–9; Martino, J., Pegg, J., & Frates, E.P. (2017). The connection prescription: using the power of social interactions and the deep desire for connectedness to empower health and wellness. *American Journal of Lifestyle Medicine*, *11*(6), 466–75; Haslam, S.A., McMahon, C., Cruwys, T., Haslam, C., Jetten, J., & Steffens, N.K. (2018). Social cure, what social cure? The propensity to underestimate the importance of social factors for health. *Social Science & Medicine*, *198*, 14–21.

9 Cohen, S., Doyle, W.J., Skoner, D.P., Rabin, B.S., & Gwaltney, J.M. (1997). Social ties and susceptibility to the common cold. *Journal of the American Medical Association*, *277*(24), 1940–4. For a discussion of these results, and more recent replications of the experiment, see: Cohen, S. (2021). Psychosocial vulnerabilities to upper respiratory infectious illness: implications for susceptibility to coronavirus disease 2019 (COVID-19). *Perspectives on Psychological Science*, *16*(1), 161–74.

10 Hemilä, H., & Chalker, E. (2013). Vitamin C for preventing and treating the common cold. *Cochrane Database of Systematic Reviews*. DOI: 10.1002/14651858.CD000980.pub4

11 Hackett, R.A., Hudson, J.L., & Chilcot, J. (2020). Loneliness and type 2 diabetes incidence: findings from the English Longitudinal Study of Ageing. *Diabetologia*, *63*, 2329–38. See also Lukaschek, K., Baumert, J., Kruse, J., Meisinger, C., & Ladwig, K.H. (2017). Sex differences in the association of social network satisfaction and the risk for type 2 diabetes. *BMC Public Health*, *17*, 1–8.

12 Kuiper, J.S., Zuidersma, M., Voshaar, R.C.O., Zuidema, S.U., van den Heuvel, E.R., Stolk, R.P., & Smidt, N. (2015). Social relationships and risk of dementia: A systematic review and meta-analysis of longitudinal cohort studies. *Ageing Research Reviews*, *22*, 39–57.

13 Valtorta, N.K., Kanaan, M., Gilbody, S., Ronzi, S., & Hanratty, B. (2016). Loneliness and social isolation as risk factors for coronary heart disease and stroke: systematic review and meta-analysis of longitudinal observational studies. *Heart*, *102*(13), 1009–16; Hakulinen, C., Pulkki-Råback, L., Virtanen, M., Jokela, M., Kivimäki, M., & Elovainio, M. (2018). Social isolation and loneliness as risk factors for myocardial infarction, stroke and mortality: UK Biobank cohort study of 479 054 men and women. *Heart*, *104*(18), 1536–42.

14 Holt-Lunstad, J., Smith, T.B., & Layton, J.B. (2010). Social relationships and mortality risk: a meta-analytic review. *PLoS Medicine*, *7*(7), e1000316.

15 Holt-Lunstad, J. (2021). The major health implications of social connection. *Current Directions in Psychological Science*, *30*(3), 251–9.

16 Snyder-Mackler, N., Burger, J.R., Gaydosh, L., Belsky, D.W., Noppert, G.A., Campos, F.A., . . . & Tung, J. (2020). Social determinants of health and survival in humans and other animals. *Science*, *368*(6493), eaax9553.

17 Eisenberger, N.I., & Cole, S.W. (2012). Social neuroscience and health: neurophysiological mechanisms linking social ties with physical health. *Nature Neuroscience*, *15*(5), 669–74; Cacioppo, J.T., Cacioppo, S., & Boomsma, D.I. (2014). Evolutionary mechanisms for loneliness. *Cognition & Emotion*, *28*(1), 3–21; Sturgeon, J.A., & Zautra, A.J. (2016). Social pain and physical pain: shared paths to resilience. *Pain Management*, *6*(1), 63–74; Zhang, M., Zhang, Y., & Kong, Y. (2019). Interaction between social pain and physical pain. *Brain Science Advances*, *5*(4), 265–73.

18 Eisenberger, N.I., Moieni, M., Inagaki, T.K., Muscatell, K.A., & Irwin, M.R. (2017). In sickness and in health: the co-regulation of inflammation and social behavior. *Neuropsychopharmacology*, *42*(1), 242–53.

19 Kim, D.A., Benjamin, E.J., Fowler, J.H., & Christakis, N.A. (2016). Social connectedness is associated with fibrinogen level in a human social network. *Proceedings of the Royal Society B: Biological Sciences*, *283*(1837), 20160958.

20 Leschak, C.J., & Eisenberger, N.I. (2019). Two distinct immune pathways linking social relationships with health: inflammatory and antiviral processes. *Psychosomatic Medicine*, *81*(8), 711; Uchino, B.N.,

Trettevik, R., Kent de Grey, R.G., Cronan, S., Hogan, J., & Baucom, B.R. (2018). Social support, social integration, and inflammatory cytokines: A meta-analysis. *Health Psychology*, *37*(5), 462.

21 For a review of all these mechanisms, see: National Academies of Sciences, Engineering, and Medicine (2020). *Social isolation and loneliness in older adults: Opportunities for the health care system*. National Academies Press.

22 The suggested existence of loneliness neurons comes from research by Ding Liu and Catherine Dulac at Harvard University. It was presented at the Society for Neuroscience meeting in 2022, but at time of writing has not yet been published in a peer-reviewed journal.

23 Uchino, B.N., & Garvey, T.S. (1997). The availability of social support reduces cardiovascular reactivity to acute psychological stress. *Journal of Behavioral Medicine*, *20*, 15–27; Heinrichs, M., Baumgartner, T., Kirschbaum, C., & Ehlert, U. (2003). Social support and oxytocin interact to suppress cortisol and subjective responses to psychosocial stress. *Biological Psychiatry*, *54*(12), 1389–98; Hooker, E.D., Campos, B., Zoccola, P.M., & Dickerson, S.S. (2018). Subjective socioeconomic status matters less when perceived social support is high: A study of cortisol responses to stress. *Social Psychological and Personality Science*, *9*(8), 981–9.

24 Hornstein, E.A., Fanselow, M.S., & Eisenberger, N.I. (2016). A safe haven: Investigating social-support figures as prepared safety stimuli. *Psychological Science*, *27*(8), 1051–60; Hornstein, E.A., Haltom, K.E., Shirole, K., & Eisenberger, N.I. (2018). A unique safety signal: Social-support figures enhance rather than protect from fear extinction. *Clinical psychological science*, *6*(3), 407–15.

25 Master, S.L., Eisenberger, N.I., Taylor, S.E., Naliboff, B.D., Shirinyan, D., & Lieberman, M.D. (2009). A picture's worth: Partner photographs reduce experimentally induced pain. *Psychological Science*, *20*(11), 1316–18; Younger, J., Aron, A., Parke, S., Chatterjee, N., & Mackey, S. (2010). Viewing pictures of a romantic partner reduces experimental pain: involvement of neural reward systems. *PloS One*, *5*(10), e13309.

26 Zalta, A.K., Tirone, V., Orlowska, D., Blais, R.K., Lofgreen, A., Klassen, B., . . . & Dent, A.L. (2021). Examining moderators of the relationship between social support and self-reported PTSD symptoms: a meta-analysis. *Psychological Bulletin*, *147*(1), 33; Hornstein, E.A., Craske, M.G., Fanselow, M.S., & Eisenberger, N.I. (2022). Reclassifying the unique inhibitory properties of social support figures: A roadmap for exploring prepared fear suppression. *Biological Psychiatry*, *91*(9), 778–85.

27 Szkody, E., Stearns, M., Stanhope, L., & McKinney, C. (2021). Stress-buffering role of social support during COVID-19. *Family Process*, *60*(3), 1002–15.

28 Ortiz-Ospina, E. (2020). Loneliness and Social Connections. Our World in Data: https://ourworldindata.org/social-connections-and-loneliness

29 Simonton, D.K. (1992). The social context of career success and course for 2,026 scientists and inventors. *Personality and Social Psychology Bulletin*, *18*(4), 452–63. More details of Simonton's methods can be found here: Simonton, D. (1984). Scientific eminence historical and contemporary: a measurement assessment. *Scientometrics*, *6*(3), 169–82.

30 Uzzi, B., & Spiro, J. (2005). Collaboration and creativity: The small world problem. *American Journal of Sociology*, *111*(2), 447–504; Uzzi, B. (2008). A social network's changing statistical properties and the quality of human innovation. *Journal of Physics A: Mathematical and Theoretical*, *41*(22), 224023; for the *West Side Story* example, see the following article: Dream teams thrive on mix of old and new blood. EurekAlert!: https://www.eurekalert.org/news-releases/621358

31 Perry-Smith, J.E. (2006). Social yet creative: the role of social relationships in facilitating individual creativity. *Academy of Management Journal*, *49*(1), 85–101; Baer, M. (2010). The strength-of-weak-ties perspective on creativity: a comprehensive examination and extension. *Journal of Applied Psychology*, *95*(3), 592.

32 Burchardi, K.B., & Hassan, T.A. (2013). The economic impact of social ties: evidence from German reunification. *Quarterly Journal of Economics*, *128*(3), 1219–71. For a discussion of these findings, see Ortiz-Ospina. Loneliness and Social Connections.

33 Meyers, L. (2007). Social relationships matter in job satisfaction. *American Psychological Association Monitor*, *38*(4), 14. Available online at: https://www.apa.org/monitor/apr07/social

34 Southwick, S.M., & Southwick, F.S. (2020). The loss of social connectedness as a major contributor to physician burnout: applying organizational and teamwork principles for prevention and recovery. *JAMA Psychiatry*, *77*(5), 449–50.

35 Haslam, S.A., McMahon, C., Cruwys, T., Haslam, C., Jetten, J., & Steffens, N.K. (2018). Social cure, what social cure?

36 Martino, J., Pegg, J., & Frates, E.P. (2017). The connection prescription.

37 Zhao, Y., Guyatt, G., Gao, Y., Hao, Q., Abdullah, R., Basmaji, J., & Foroutan, F. (2022). Living alone and all-cause mortality in community-dwelling adults: A systematic review and meta-analysis. *EClinicalMedicine*, *54*, 101677; Stavrova, O., & Ren, D. (2021). Is more always better?

Examining the nonlinear association of social contact frequency with physical health and longevity. *Social Psychological and Personality Science, 12*(6), 1058–70.

38 Campo, R.A., Uchino, B.N., Holt-Lunstad, J., Vaughn, A., Reblin, M., & Smith, T.W. (2009). The assessment of positivity and negativity in social networks: the reliability and validity of the social relationships index. *Journal of Community Psychology, 37*(4), 471–86.

39 Holt-Lunstad, J., Uchino, B.N., Smith, T.W., Olson-Cerny, C., & Nealey-Moore, J.B. (2003). Social relationships and ambulatory blood pressure: structural and qualitative predictors of cardiovascular function during everyday social interactions. *Health Psychology, 22*(4), 388.

40 Holt-Lunstad, J., & Clark B.D. (2014). Social stressors and cardiovascular response: Influence of ambivalent relationships and behavioral ambivalence. *International Journal of Psychophysiology, 93*, 381–9.

41 Carlisle, M., Uchino, B.N., Sanbonmatsu, D.M., Smith, T.W., Cribbet, M.R., Birmingham, W., . . . & Vaughn, A.A. (2012). Subliminal activation of social ties moderates cardiovascular reactivity during acute stress. *Health Psychology, 31*(2), 217.

42 Ross, K.M., Rook, K., Winczewski, L., Collins, N., & Dunkel Schetter, C. (2019). Close relationships and health: The interactive effect of positive and negative aspects. *Social and Personality Psychology Compass, 13*(6), e12468.

43 Holt-Lunstad, J., & Uchino, B.N. (2019). Social ambivalence and disease (SAD): a theoretical model aimed at understanding the health implications of ambivalent relationships. *Perspectives on Psychological Science, 14*(6), 941–66.

44 Herr, R.M., Birmingham, W.C., van Harreveld, F., van Vianen, A.E., Fischer, J.E., & Bosch, J.A. (2022). The relationship between ambivalence towards supervisor's behavior and employee's mental health. *Scientific Reports, 12*(1), 9555.

45 Retrieved from Ghinsberg's website on 10 February 2023: https://ghinsberg.com/blog/about-me-my-vision-my-mission

46 Ghinsberg, Y. (1993). *Back to Tuichi* (p. 135). Random House.

CHAPTER 2: *How We Connect*

1 Plath, S. (1975). *Letters Home* (pp. 46–8). Faber & Faber.

2 Plath, S. (2000). *The Unabridged Journals of Sylvia Plath: 1950–1962* (pp. 28–31). Anchor.

3 As with many scientific concepts, shared reality can be defined in various ways. In this book, I lean on the definition adopted by Gerald Echterhoff, E. Tory Higgins and John Levine: Echterhoff, G., Higgins, E.T., & Levine, J.M. (2009). Shared reality: experiencing commonality

with others' inner states about the world. *Perspectives on Psychological Science*, 4(5), 496–521.

4 Konstan, D. (2018). 'One Soul in Two Bodies: Distributed Cognition and Ancient Greek Friendship'. In *Distributed Cognition in Classical Antiquity*, Edinburgh University Press, pp. 209–24. https://doi.org/10.1515/9781474429764-014

5 Pinel, E.C., Long, A.E., Landau, M.J., Alexander, K., & Pyszczynski, T. (2006). Seeing I to I: a pathway to interpersonal connectedness. *Journal of Personality and Social Psychology*, 90(2), 243; Higgins, E. Tory. *Shared Reality* (pp. 251–2). Oxford University Press. Kindle Edition.

6 Pinel, E.C., & Long, A.E. (2012). When I's meet: sharing subjective experience with someone from the outgroup. *Personality and Social Psychology Bulletin*, 38(3), 296–307; Pinel, E.C., Long, A.E., & Crimin, L.A. (2008). We're warmer (they're more competent): I-sharing and African-Americans' perceptions of the ingroup and outgroup. *European Journal of Social Psychology*, 38(7), 1184–92.

7 Pinel, E.C., Fuchs, N.A., & Benjamin, S. (2022). I-sharing across the aisle: can shared subjective experience bridge the political divide? *Journal of Applied Social Psychology*, 52(6), 407–13. The inkblot question first appeared in Pinel, E.C., Long, A.E., & Huneke, M. (2015). In the blink of an I: On delayed but identical subjective reactions and their effect on self-interested behavior. *The Journal of Social Psychology*, 155(6), 605–16.

8 Huneke, M., & Pinel, E.C. (2016). Fostering selflessness through I-sharing. *Journal of Experimental Social Psychology*, 63, 10–18.

9 Pinel, E. C., Long, A. E., Landau, M. J., Alexander, K., & Pyszczynski, T. (2006). Seeing I to I.

10 Rivera, G.N., Smith, C.M., & Schlegel, R.J. (2019). A window to the true self: the importance of I-sharing in romantic relationships. *Journal of Social and Personal Relationships*, 36(6), 1640–50.

11 Rossignac-Milon, M., Bolger, N., Zee, K.S., Boothby, E.J., & Higgins, E.T. (2021). Merged minds: generalized shared reality in dyadic relationships. *Journal of Personality and Social Psychology*, 120(4), 882.

12 The diagram used here has been adapted by the author, after Aron, A., Aron, E.N., & Smollan, D. (1992). Inclusion of other in the self scale and the structure of interpersonal closeness. *Journal of Personality and Social Psychology*, 63(4), 596–612.

13 This interpretation of a shared stream of consciousness can be found here: Higgins, E.T., Rossignac-Milon, M., & Echterhoff, G. (2021). Shared reality: from sharing-is-believing to merging minds. *Current Directions in Psychological Science*, 30(2), 103–10.

14 Montaigne, Michel de. *On Friendship* (p. 11). Penguin Great Ideas. Kindle Edition.

15 Montaigne, Michel de. *On Friendship* (p. 8).
16 Parkinson, C., Kleinbaum, A.M., & Wheatley, T. (2018). Similar neural responses predict friendship. *Nature Communications*, 9(1), 1–14. A further analysis of the data confirmed these conclusions: Hyon, R., Kleinbaum, A.M., & Parkinson, C. (2020). Social network proximity predicts similar trajectories of psychological states: evidence from multi-voxel spatiotemporal dynamics. *NeuroImage*, *216*, 116492.
17 For an example of one of these studies, see: Kinreich, S., Djalovski, A., Kraus, L., Louzoun, Y., & Feldman, R. (2017). Brain-to-brain synchrony during naturalistic social interactions. *Scientific Reports*, 7(1), 17060. For a review of this field of research, see: Baek, E.C., & Parkinson, C. (2022). Shared understanding and social connection: integrating approaches from social psychology, social network analysis, and neuroscience. *Social and Personality Psychology Compass*, *16*(11), e12710.
18 Nummenmaa, L., Lahnakoski, J.M., & Glerean, E. (2018). Sharing the social world via intersubject neural synchronisation. *Current Opinion in Psychology*, *24*, 7–14.
19 Luft, C.D.B., Zioga, I., Giannopoulos, A., Di Bona, G., Binetti, N., Civilini, A., . . . & Mareschal, I. (2022). Social synchronization of brain activity increases during eye-contact. *Communications Biology*, *5*(1), 1–15.
20 Mu, Y., Cerritos, C., & Khan, F. (2018). Neural mechanisms underlying interpersonal coordination: a review of hyperscanning research. *Social and Personality Psychology Compass*, *12*(11), e12421.
21 Oishi, S., Krochik, M., & Akimoto, S. (2010). Felt understanding as a bridge between close relationships and subjective well-being: antecedents and consequences across individuals and cultures. *Social and Personality Psychology Compass*, 4(6), 403–16.
22 Selcuk, E., Gunaydin, G., Ong, A.D., & Almeida, D.M. (2016). Does partner responsiveness predict hedonic and eudaimonic well-being? A 10-year longitudinal study. *Journal of Marriage and Family*, 78(2), 311–25; Helm, P.J., Medrano, M.R., Allen, J.J., & Greenberg, J. (2020). Existential isolation, loneliness, depression, and suicide ideation in young adults. *Journal of Social and Clinical Psychology*, 39(8), 641–74; Constantino, M.J., Sommer, R.K., Goodwin, B.J., Coyne, A.E., & Pinel, E.C. (2019). Existential isolation as a correlate of clinical distress, beliefs about psychotherapy, and experiences with mental health treatment. *Journal of Psychotherapy Integration*, 29(4), 389.
23 Shamay-Tsoory, S.G., Saporta, N., Marton-Alper, I.Z., & Gvirts, H.Z.

(2019). Herding brains: a core neural mechanism for social alignment. *Trends in Cognitive Sciences*, *23*(3), 174–86.

24 Durkheim, E. (1965). *The Elementary Forms of the Religious Life* (J.W. Swain, Trans.). Free Press. The example of firewalking comes from the following paper: Konvalinka, I., Xygalatas, D., Bulbulia, J., Schjødt, U., Jegindø, E.M., Wallot, S., . . . & Roepstorff, A. (2011). Synchronized arousal between performers and related spectators in a fire-walking ritual. *Proceedings of the National Academy of Sciences*, *108*(20), 8514–19.

25 Wheatley, T., Kang, O., Parkinson, C., & Looser, C.E. (2012). From mind perception to mental connection: Synchrony as a mechanism for social understanding. *Social and Personality Psychology Compass*, *6*(8), 589–606.

26 Wiltermuth, S.S., & Heath, C. (2009). Synchrony and cooperation. *Psychological Science*, *20*(1), 1–5.

27 Miles, L.K., Nind, L.K., Henderson, Z., & Macrae, C.N. (2010). Moving memories: Behavioral synchrony and memory for self and others. *Journal of Experimental Social Psychology*, *46*(2), 457–60; Valdesolo, P., Ouyang, J., & DeSteno, D. (2010). The rhythm of joint action: Synchrony promotes cooperative ability. *Journal of experimental social psychology*, *46*(4), 693–95; Tarr, B., Launay, J., & Dunbar, R.I. (2014). Music and social bonding: 'self-other' merging and neurohormonal mechanisms. *Frontiers in Psychology*, *5*, 1096.

28 Tarr, B., Launay, J., & Dunbar, R.I. (2014). Music and social bonding; Savage, P.E., Loui, P., Tarr, B., Schachner, A., Glowacki, L., Mithen, S., & Fitch, W.T. (2021). Music as a coevolved system for social bonding. *Behavioral and Brain Sciences*, *44*, e59.

29 Smith, P. (2015). M *Train* (p. 87). Knopf.

30 Aron, A., Lewandowski Jr, G.W., Mashek, D., & Aron, E.N. (2013). The self-expansion model of motivation and cognition in close relationships. *The Oxford Handbook of Close Relationships*, 90–115; Aron, A., Lewandowski, G., Branand, B., Mashek, D., & Aron, E. (2022). Self-expansion motivation and inclusion of others in self: an updated review. *Journal of Social and Personal Relationships*, *39*(12), 3821–52.

31 Sparks, J., Daly, C., Wilkey, B.M., Molden, D.C., Finkel, E.J., & Eastwick, P.W. (2020). Negligible evidence that people desire partners who uniquely fit their ideals. *Journal of Experimental Social Psychology*, *90*, 103968.

32 Huang, S.A., Ledgerwood, A., & Eastwick, P.W. (2020). How do ideal friend preferences and interaction context affect friendship formation? Evidence for a domain-general relationship initiation process. *Social Psychological and Personality Science*, *11*(2), 226–35.

33 These paragraphs include insights inspired by my conversations with Harry Reiss and Paul Eastwick for the following article: 'A sexual destiny mindset' – and the other red flags of romantic chemistry. *Guardian*: https://www.theguardian.com/science/2023/feb/12/the-science-of-romantic-chemistry-and-those-not-so-obvious-red-flags. See also Reis, H.T., Regan, A., & Lyubomirsky, S. (2022). Interpersonal chemistry: what is it, how does it emerge, and how does it operate? *Perspectives on Psychological Science*, *17*(2), 530–58.

34 Aron, A., Steele, J.L., Kashdan, T.B., & Perez, M. (2006). When similars do not attract: Tests of a prediction from the self-expansion model. *Personal Relationships*, *13*(4), 387–96; Aron, A., Lewandowski, G., Branand, B., Mashek, D., & Aron, E. (2022). Self-expansion motivation and inclusion of others in self: an updated review; Santucci, K., Khullar, T.H., & Dirks, M.A. (2022). Through thick and thin?: Young adults' implicit beliefs about friendship and their reported use of dissolution and maintenance strategies with same-gender friends. *Social Development*, *31*(2), 480–96.

35 Cirelli, L.K., Wan, S.J., & Trainor, L.J. (2014). Fourteen-month-old infants use interpersonal synchrony as a cue to direct helpfulness. *Philosophical Transactions of the Royal Society B: Biological Sciences*, *369*(1658), 20130400.

36 Göritz, A.S., & Rennung, M. (2019). Interpersonal synchrony increases social cohesion, reduces work-related stress and prevents sickdays: a longitudinal field experiment. *Gruppe. Interaktion. Organisation: Zeitschrift für angewandte Organisationspsychologie*, *50*, 83–94; see also Hu, Y., Cheng, X., Pan, Y., & Hu, Y. (2022). The intrapersonal and interpersonal consequences of interpersonal synchrony. *Acta Psychologica*, *224*, 103513.

37 Baranowski-Pinto, G., Profeta, V.L.S., Newson, M., Whitehouse, H., & Xygalatas, D. (2022). Being in a crowd bonds people via physiological synchrony. *Scientific Reports*, *12*(1), 1–10.

38 Bastian, B., Jetten, J., & Ferris, L.J. (2014). Pain as social glue: shared pain increases cooperation. *Psychological Science*, *25*(11), 2079–85; Peng, W., Lou, W., Huang, X., Ye, Q., Tong, R.K.Y., & Cui, F. (2021). Suffer together, bond together: brain-to-brain synchronization and mutual affective empathy when sharing painful experiences. *NeuroImage*, *238*, 118249.

39 Aron, A., Lewandowski, G., Branand, B., Mashek, D., & Aron, E. (2022). Self-expansion motivation and inclusion of others in self: an updated review.

40 Nin, A. (1974). *The Journals of Anaïs Nin: Volume 2 (1934–1939)* (p. 202). Quartet.

CHAPTER 3: *The Personality Myth*

1 Corti, K., & Gillespie, A. (2016). Co-constructing intersubjectivity with artificial conversational agents: people are more likely to initiate repairs of misunderstandings with agents represented as human. *Computers in Human Behavior, 58*, 431–42.

2 Epley, N., & Schroeder, J. (2014). Mistakenly seeking solitude. *Journal of Experimental Psychology: General, 143*(5), 1980.

3 Schroeder, J., Lyons, D., & Epley, N. (2022). Hello, stranger? Pleasant conversations are preceded by concerns about starting one. *Journal of Experimental Psychology: General, 151*(5), 1141.

4 Sandstrom, G.M., & Dunn, E.W. (2014). Is efficiency overrated? Minimal social interactions lead to belonging and positive affect. *Social Psychological and Personality Science, 5*(4), 437–42.

5 Sandstrom, G.M., & Boothby, E.J. (2021). Why do people avoid talking to strangers? A mini meta-analysis of predicted fears and actual experiences talking to a stranger. *Self and Identity*, 20(1), 47–71.

6 Boothby, E.J., Cooney, G., Sandstrom, G.M., & Clark, M.S. (2018). The liking gap in conversations: do people like us more than we think? *Psychological Science, 29*(11), 1742–56.

7 Mastroianni, A.M., Cooney, G., Boothby, E.J., & Reece, A.G. (2021). The liking gap in groups and teams. *Organizational Behavior and Human Decision Processes, 162*, 109–22.

8 Wolf, W., Nafe, A., & Tomasello, M. (2021). The development of the liking gap: children older than 5 years think that partners evaluate them less positively than they evaluate their partners. *Psychological Science, 32*(5), 789–98.

9 Savitsky, K., Epley, N., & Gilovich, T. (2001). Do others judge us as harshly as we think? Overestimating the impact of our failures, shortcomings, and mishaps. *Journal of Personality and Social Psychology, 81*(1), 44; Gilovich, T., Medvec, V.H., & Savitsky, K. (2000). The spotlight effect in social judgment: an egocentric bias in estimates of the salience of one's own actions and appearance. *Journal of Personality and Social Psychology, 78*(2), 211.

10 De Jong, P.J., & Dijk, C. (2013). Social effects of facial blushing: influence of context and actor versus observer perspective. *Social and Personality Psychology Compass, 7*(1), 13–26; Thorstenson, C.A., Pazda, A.D., & Lichtenfeld, S. (2020). Facial blushing influences perceived embarrassment and related social functional evaluations. *Cognition and Emotion, 34*(3), 413–26.

11 Whitehouse, J., Milward, S.J., Parker, M.O., Kavanagh, E., & Waller,

B.M. (2022). Signal value of stress behaviour. *Evolution and Human Behavior*, *43*(4), 325–33.

12 Zell, E., Strickhouser, J.E., Sedikides, C., & Alicke, M.D. (2020). The better-than-average effect in comparative self-evaluation: A comprehensive review and meta-analysis. *Psychological Bulletin*, *146*(2), 118.

13 Elsaadawy, N., & Carlson, E.N. (2022). Do you make a better or worse impression than you think? *Journal of Personality and Social Psychology*, *123*(6), 1407–20.

14 Welker, C., Walker, J., Boothby, E., & Gilovich, T. (2023). Pessimistic assessments of ability in informal conversation. *Journal of Applied Social Psychology*, 53, 555–69; Atir, S., Zhao, X., & Echelbarger, M. (2023). Talking to strangers: Intention, competence, and opportunity. *Current Opinion in Psychology*, 101588.

15 This saying is often attributed to Maya Angelou, but seems to have been in circulation decades before she started writing her memoirs: https://quoteinvestigator.com/2014/04/06/they-feel/

16 Sandstrom, G.M., Boothby, E.J., & Cooney, G. (2022). Talking to strangers: a week-long intervention reduces psychological barriers to social connection. *Journal of Experimental Social Psychology*, *102*, 104356.

17 Rollings, J., Micheletta, J., Van Laar, D., & Waller, B.M. (2023). Personality traits predict social network size in older adults. *Personality and Social Psychology Bulletin*, *49*(6), 925–38; Gale, C.R., Booth, T., Mõttus, R., Kuh, D., & Deary, I.J. (2013). Neuroticism and Extraversion in youth predict mental wellbeing and life satisfaction 40 years later. *Journal of Research in Personality*, 47(6), 687–97; Rizzuto, D., Mossello, E., Fratiglioni, L., Santoni, G., & Wang, H.X. (2017). Personality and survival in older age: the role of lifestyle behaviors and health status. *The American Journal of Geriatric Psychiatry*, *25*(12), 1363–72.

18 Zelenski, J.M., Whelan, D.C., Nealis, L.J., Besner, C.M., Santoro, M.S., & Wynn, J.E. (2013). Personality and affective forecasting: Trait introverts underpredict the hedonic benefits of acting extraverted. *Journal of Personality and Social Psychology*, *104*(6), 1092.

19 Margolis, S., & Lyubomirsky, S. (2020). Experimental manipulation of extraverted and introverted behavior and its effects on well-being. *Journal of Experimental Psychology: General*, *149*(4), 719.

20 Duffy, K.A., Helzer, E.G., Hoyle, R.H., Fukukura Helzer, J., & Chartrand, T.L. (2018). Pessimistic expectations and poorer experiences: the role of (low) extraversion in anticipated and experienced enjoyment of social interaction. *PloS One*, *13*(7), e0199146; Zelenski, J.M., Whelan,

D.C., Nealis, L.J., Besner, C.M., Santoro, M.S., & Wynn, J.E. (2013). Personality and affective forecasting.

The following paper offers a discussion of these results in the broader context of social behaviour and its benefits: Epley, N., Kardas, M., Zhao, X., Atir, S., & Schroeder, J. (2022). Undersociality: miscalibrated social cognition can inhibit social connection. *Trends in Cognitive Sciences*, *26*(5), 406–18.

21 Hudson, N.W., & Fraley, R.C. (2015). Volitional personality trait change: can people choose to change their personality traits? *Journal of Personality and Social Psychology*, *109*(3), 490; Stieger, M., Flückiger, C., Rüegger, D., Kowatsch, T., Roberts, B.W., & Allemand, M. (2021). Changing personality traits with the help of a digital personality change intervention. *Proceedings of the National Academy of Sciences*, *118*(8), e2017548118.

22 Kivity, Y., & Huppert, J.D. (2016). Does cognitive reappraisal reduce anxiety? A daily diary study of a micro-intervention with individuals with high social anxiety. *Journal of Consulting and Clinical Psychology*, *84*(3), 269; Duijndam, S., Karreman, A., Denollet, J., & Kupper, N. (2020). Emotion regulation in social interaction: physiological and emotional responses associated with social inhibition. *International Journal of Psychophysiology*, *158*, 62–72.

23 Savitsky, K., Epley, N., & Gilovich, T. (2001). Do others judge us as harshly as we think?

24 You can find out more about this strategy at the Greater Good Science Centre website, a science-based initiative from the University of California, Berkeley: https://greatergood.berkeley.edu/article/item/how_to_be_yourself_when_you_have_social_anxiety

25 Wax, R. (2016). *A Mindfulness Guide for the Frazzled* (p. 41). Penguin Random House UK.

26 Breines, J.G., & Chen, S. (2012). Self-compassion increases self-improvement motivation. *Personality and Social Psychology Bulletin*, *38*(9), 1133–43; Vazeou-Nieuwenhuis, A., & Schumann, K. (2018). Self-compassionate and apologetic? How and why having compassion toward the self relates to a willingness to apologize. *Personality and Individual Differences*, *124*, 71–6.

27 Werner, K.H., Jazaieri, H., Goldin, P.R., Ziv, M., Heimberg, R.G., & Gross, J.J. (2012). Self-compassion and social anxiety disorder. *Anxiety, Stress & Coping*, *25*(5), 543–58.

28 Ketay, S., Beck, L.A., & Dajci, J. (2022). Self-compassion and social stress: links with subjective stress and cortisol responses. *Self and Identity*, 1–20.

29 You can take the full test at https://self-compassion.org/self-compassion-test/

30 Stevenson, J., Mattiske, J.K., & Nixon, R.D. (2019). The effect of a brief online self-compassion versus cognitive restructuring intervention on trait social anxiety. *Behaviour Research and Therapy*, *123*, 103492; Teale Sapach, M.J., & Carleton, R.N. (2023). Self-compassion training for individuals with social anxiety disorder: a preliminary randomized controlled trial. *Cognitive Behaviour Therapy*, *52*(1), 18–37.

31 Burnette, J.L., Knouse, L.E., Vavra, D.T., O'Boyle, E., & Brooks, M.A. (2020). Growth mindsets and psychological distress: A meta-analysis. *Clinical Psychology Review*, 77, 101816. See also: Schroder, H.S., Kneeland, E.T., Silverman, A.L., Beard, C., & Björgvinsson, T. (2019). Beliefs about the malleability of anxiety and general emotions and their relation to treatment outcomes in acute psychiatric treatment. *Cognitive Therapy and Research*, *43*(2), 312–23; see also Schleider, J., & Weisz, J. (2018). A single-session growth mindset intervention for adolescent anxiety and depression: 9-month outcomes of a randomized trial. *Journal of Child Psychology and Psychiatry*, *59*(2), 160–70.

32 Valentiner, D.P., Mounts, N.S., Durik, A.M., & Gier-Lonsway, S.L. (2011). Shyness mindset: applying mindset theory to the domain of inhibited social behavior. *Personality and Individual Differences*, *50*(8), 1174–9.

33 Valentiner, D.P., Jencius, S., Jarek, E., Gier-Lonsway, S.L., & McGrath, P.B. (2013). Pre-treatment shyness mindset predicts less reduction of social anxiety during exposure therapy. *Journal of Anxiety Disorders*, *27*(3), 267–71.

34 You can read more about this research in my book *The Expectation Effect* (2022). For a recent, large study demonstrating the benefits, see: Yeager, D.S., Bryan, C.J., Gross, J.J., Murray, J.S., Krettek Cobb, D., Santos, P.H.F., . . . & Jamieson, J.P. (2022). A synergistic mindsets intervention protects adolescents from stress. *Nature*, 607(7919), 512–20.

CHAPTER 4: *Overcoming Egocentric Thinking*

1 In popular retellings, this tale is sometimes exaggerated, but the basic details of the misunderstanding have been confirmed by Hoover's assistant: Mikkelson, B. (1999). Watch the Borders!. Snopes: https://www.snopes.com/fact-check/watch-the-borders.

2 Hoover's behaviour appears as an example in: Keysar, B., & Barr, D.J. (2002). Self-anchoring in conversation: why language users do not do what they 'should'. In Gilovich, T., Griffin, D., & Kahneman, D. (eds), *Heuristics and Biases: The Psychology of Intuitive Judgment* (pp. 150–66). Cambridge University Press.

3 When it is presented less obviously, many adults even fail the Sally-Anne test: Birch, S.A., & Bloom, P. (2007). The curse of knowledge in reasoning about false beliefs. *Psychological Science, 18*(5), 382–6.
4 Keysar, B., Barr, D.J., Balin, J.A., & Brauner, J.S. (2000). Taking perspective in conversation: The role of mutual knowledge in comprehension. *Psychological Science, 11*(1), 32–8.
5 Epley, N., Morewedge, C.K., & Keysar, B. (2004). Perspective taking in children and adults: equivalent egocentrism but differential correction. *Journal of Experimental Social Psychology*, 40(6), 760–8.
6 Epley, N., Keysar, B., Van Boven, L., & Gilovich, T. (2004). Perspective taking as egocentric anchoring and adjustment. *Journal of Personality and Social Psychology*, 87(3), 327; Lin, S., Keysar, B., & Epley, N. (2010). Reflexively mindblind: using theory of mind to interpret behavior requires effortful attention. *Journal of Experimental Social Psychology*, 46(3), 551–6.
7 Krueger, J., & Clement, R.W. (1994). The truly false consensus effect: an ineradicable and egocentric bias in social perception. *Journal of Personality and Social Psychology*, 67(4), 596.
8 Dunn, M., Thomas, J.O., Swift, W., & Burns, L. (2012). Elite athletes' estimates of the prevalence of illicit drug use: evidence for the false consensus effect. *Drug and Alcohol Review*, 31(1), 27–32.
9 Gilovich, T., Savitsky, K., & Medvec, V.H. (1998). The illusion of transparency: biased assessments of others' ability to read one's emotional states. *Journal of Personality and Social Psychology*, 75(2), 332.
10 Savitsky, K., & Gilovich, T. (2003). The illusion of transparency and the alleviation of speech anxiety. *Journal of Experimental Social Psychology*, 39(6), 618–25.
11 Keysar, B. (1994). The illusory transparency of intention: linguistic perspective taking in text. *Cognitive Psychology*, 26(2), 165–208.
12 Kruger, J., Epley, N., Parker, J., & Ng, Z.W. (2005). Egocentrism over e-mail: can we communicate as well as we think? *Journal of Personality and Social Psychology*, 89(6), 925.
13 Keysar, B., & Henly, A.S. (2002). Speakers' overestimation of their effectiveness. *Psychological Science*, 13(3), 207–12; see also Keysar, B., & Barr, D.J. (2002). Self-anchoring in conversation: why language users do not do what they 'should'. In Gilovich, T., Griffin, D., & Kahneman, D. (eds), *Heuristics and Biases: The Psychology of Intuitive Judgment*, (pp. 150–66).
14 Savitsky, K., Keysar, B., Epley, N., Carter, T., & Swanson, A. (2011). The closeness-communication bias: increased egocentrism among friends versus strangers. *Journal of Experimental Social Psychology*, 47(1), 269–73.

15 Cheung, H. (2019). YouGov survey: British sarcasm 'lost on Americans'. BBC: https://www.bbc.com/news/world-us-canada-46846467
16 Lau, B.K.Y., Geipel, J., Wu, Y., & Keysar, B. (2022). The extreme illusion of understanding. *Journal of Experimental Psychology: General, 151*(11), 2957–62.
17 Chang, V.Y., Arora, V.M., Lev-Ari, S., D'Arcy, M., & Keysar, B. (2010). Interns overestimate the effectiveness of their hand-off communication. *Pediatrics, 125*(3), 491–6.
18 Mishap Investigation Board (1999). Phase I Report. https://llis.nasa.gov/llis_lib/pdf/1009464main1_0641-mr.pdf
19 Savitsky, K., Keysar, B., Epley, N., Carter, T., & Swanson, A. The closeness-communication bias.
20 Eyal, T., Steffel, M., & Epley, N. (2018). Perspective mistaking: accurately understanding the mind of another requires getting perspective, not taking perspective. *Journal of Personality and Social Psychology, 114*(4), 547.
21 Goldstein, N.J., Vezich, I.S., & Shapiro, J.R. (2014). Perceived perspective taking: when others walk in our shoes. *Journal of Personality and Social Psychology, 106*(6), 941.
22 Edwards, R., Bybee, B.T., Frost, J.K., Harvey, A.J., & Navarro, M. (2017). That's not what I meant: how misunderstanding is related to channel and perspective-taking. *Journal of Language and Social Psychology, 36*(2), 188–210.
23 Cahill, V.A., Malouff, J.M., Little, C.W., & Schutte, N.S. (2020). Trait perspective taking and romantic relationship satisfaction: a meta-analysis. *Journal of Family Psychology, 34*(8), 1025.
24 Eyal, T., Steffel, M., & Epley, N. (2018). Perspective mistaking.
25 Damen, D., van Amelsvoort, M., van der Wijst, P., Pollmann, M., & Krahmer, E. (2021). Lifting the curse of knowing: how feedback improves perspective-taking. *Quarterly Journal of Experimental Psychology, 74*(6), 1054–69.

CHAPTER 5: *The Art of Conversation*

1 West, R. (1982). *The Harsh Voice* (p. 63). London: Virago. (Originally published 1929).
2 Hazlitt, W. (1870). *The Plain Speaker* (pp. 50–1). London: Bell and Daldy. (The essay first appeared in the *London Magazine*, 1820).
3 Huang, K., Yeomans, M., Brooks, A.W., Minson, J., & Gino, F. (2017). It doesn't hurt to ask: question-asking increases liking. *Journal of Personality and Social Psychology, 113*(3), 430. For a lengthier discussion of the methods see: https://dash.harvard.edu/bitstream/

handle/1/35647952/huangyeomansbrooksminsongino_QuestionAsking_Manuscript.pdf. The term 'boomerasking' can be found in the following paper: Yeomans, M., Schweitzer, M.E., & Brooks, A.W. (2022). The conversational circumplex: identifying, prioritizing, and pursuing informational and relational motives in conversation. *Current Opinion in Psychology*, *44*, 293–302.

4 This research, by Alison Wood Brooks, Michael Yeomans and Michael Norton, was presented at the International Association for Conflict Management conference in 2023.

5 Zhou, J., & Fredrickson, B.L. (2023). Listen to Resonate: Better Listening as a Gateway to Interpersonal Positivity Resonance through Enhanced Sensory Connection and Perceived Safety. *Current Opinion in Psychology*, 101669; Lloyd, K.J., Boer, D., Keller, J.W., & Voelpel, S. (2015). Is my boss really listening to me? The impact of perceived supervisor listening on emotional exhaustion, turnover intention, and organizational citizenship behavior. *Journal of Business Ethics*, *130*, 509–24.

6 Castro, D.R., Anseel, F., Kluger, A.N., Lloyd, K.J., & Turjeman-Levi, Y. (2018). Mere listening effect on creativity and the mediating role of psychological safety. *Psychology of Aesthetics, Creativity, and the Arts*, *12*(4), 489.

7 Collins, H.K. (2022). When listening is spoken. *Current Opinion in Psychology*, 101402.

8 This example is described in the following paper: Itzchakov, G., Reis, H.T., & Weinstein, N. (2022). How to foster perceived partner responsiveness: high-quality listening is key. *Social and Personality Psychology Compass*, *16*(1), e12648.

9 Misra, S., Cheng, L., Genevie, J., & Yuan, M. (2016). The iPhone effect: the quality of in-person social interactions in the presence of mobile devices. *Environment and Behavior*, *48*(2), 275–98.

10 Dwyer, R.J., Kushlev, K., & Dunn, E.W. (2018). Smartphone use undermines enjoyment of face-to-face social interactions. *Journal of Experimental Social Psychology*, *78*, 233–9. For further evidence, see: Roberts, J.A., & David, M.E. (2022). Partner phubbing and relationship satisfaction through the lens of social allergy theory. *Personality and Individual Differences*, *195*, 111676; Al Saggaf, Y., & O'Donnell, S.B. (2019). Phubbing: perceptions, reasons behind, predictors, and impacts. *Human Behavior and Emerging Technologies*, *1*(2), 132–40.

11 Rossignac-Milon, M. (2019). *Merged Minds: Generalized Shared Reality in Interpersonal Relationships*. Columbia University.

12 McFarland, D.A., Jurafsky, D., & Rawlings, C. (2013). Making the connection: social bonding in courtship situations. *American Journal of Sociology*, *118*(6), 1596–649.

13 Mein, C., Fay, N., & Page, A.C. (2016). Deficits in joint action explain why socially anxious individuals are less well liked. *Journal of Behavior Therapy and Experimental Psychiatry*, *50*, 147–51; Günak, M.M., Clark, D.M., & Lommen, M.J. (2020). Disrupted joint action accounts for reduced likability of socially anxious individuals. *Journal of Behavior Therapy and Experimental Psychiatry*, *68*, 101512.

14 Flynn, F.J., Collins, H., & Zlatev, J. (2022). Are you listening to me? The negative link between extraversion and perceived listening. *Personality and Social Psychology Bulletin*, 01461672211072815.

15 Hirschi, Q., Wilson, T.D., & Gilbert, D.T. (2022). Speak up! Mistaken beliefs about how much to talk in conversations. *Personality and Social Psychology Bulletin*, 01461672221104927.

16 Aron, A., Melinat, E., Aron, E.N., Vallone, R.D., & Bator, R.J. (1997). The experimental generation of interpersonal closeness: a procedure and some preliminary findings. *Personality and Social Psychology Bulletin*, *23*(4), 363–77.

17 The following provides a replication and confirmation that the fast friends procedure is superior to unstructured conversation, and that it works equally well in video calls and face-to-face interactions: Sprecher, S. (2021). Closeness and other affiliative outcomes generated from the fast friends procedure: a comparison with a small-talk task and unstructured self-disclosure and the moderating role of mode of communication. *Journal of Social and Personal Relationships*, *38*(5), 1452–71.

18 Stürmer, S., Ihme, T.A., Fisseler, B., Sonnenberg, K., & Barbarino, M.L. (2018). Promises of structured relationship building for higher distance education: evaluating the effects of a virtual fast-friendship procedure. *Computers & Education*, *124*, 51–61.

19 Lytle, A., & Levy, S.R. (2015). Reducing heterosexuals' prejudice toward gay men and lesbian women via an induced cross-orientation friendship. *Psychology of Sexual Orientation and Gender Diversity*, *2*(4), 447.

20 Echols, L., & Ivanich, J. (2021). From 'fast friends' to true friends: can a contact intervention promote friendships in middle school? *Journal of Research on Adolescence*, *31*(4), 1152–71.

21 Kardas, M., Kumar, A., & Epley, N. (2022). Overly shallow?: miscalibrated expectations create a barrier to deeper conversation. *Journal of Personality and Social Psychology*, *122*(3), 367.

22 Thorson, K.R., Ketay, S., Roy, A.R., & Welker, K.M. (2021). Self-disclosure is associated with adrenocortical attunement between new acquaintances. *Psychoneuroendocrinology*, *132*, 105323.

23 Inagaki, T.K. (2018). Opioids and social connection. *Current Directions in Psychological Science*, *27*(2), 85–90.

24 Slatcher, R.B. (2010). When Harry and Sally met Dick and Jane: Creating closeness between couples. *Personal Relationships*, *17*(2), 279–97; Welker, K.M., Baker, L., Padilla, A., Holmes, H., Aron, A., & Slatcher, R.B. (2014). Effects of self-disclosure and responsiveness between couples on passionate love within couples. *Personal Relationships*, *21*(4), 692–708.

25 Milek, A., Butler, E.A., Tackman, A.M., Kaplan, D.M., Raison, C.L., Sbarra, D.A., . . . & Mehl, M.R. (2018). 'Eavesdropping on happiness' revisited: a pooled, multisample replication of the association between life satisfaction and observed daily conversation quantity and quality. *Psychological Science*, *29*(9), 1451–62.

26 Sanchez, K.L., Kalkstein, D.A., & Walton, G.M. (2022). A threatening opportunity: The prospect of conversations about race-related experiences between Black and White friends. *Journal of Personality and Social Psychology*, *122*(5), 853.

27 Reczek, C.E., Reczek, R., & Bosley-Smith, E. (2022). *Families We Keep: LGBTQ People and Their Enduring Bonds with Parents*. NYU Press.

28 Sanchez, K.L., Kalkstein, D.A., & Walton, G.M. (2022). A threatening opportunity.

29 Cooney, G., Gilbert, D.T., & Wilson, T.D. (2017). The novelty penalty: why do people like talking about new experiences but hearing about old ones? *Psychological Science*, *28*(3), 380–94.

30 Hirschi, Q., Wilson, T.D., & Gilbert, D.T. Speak up! Mistaken beliefs about how much to talk in conversations.

CHAPTER 6: *Expressing Appreciation*

1 This version of 'The Fox and the Crow' appears in *The Aesop for Children: with Pictures by Milo Winter*, published by Rand, McNally & Co in 1919. It is considered to be in the public domain in the US and can be accessed online from the Library of Congress: https://read.gov/aesop/about.html

2 La Fontaine, J. (2010). *The Complete Fables of Jean de la Fontaine* (Shapiro, N., Trans.) (p. 5). University of Illinois Press.

3 Alighieri, D. (2006). *Inferno* (Kirkpatrick, R., Trans) (p. 158). Penguin UK.

4 Bareket-Bojmel, L., Hochman, G., & Ariely, D. (2017). It's (not) all about the Jacksons: testing different types of short-term bonuses in the field. *Journal of Management*, *43*(2), 534–54.

5 Bradler, C., Dur, R., Neckermann, S., & Non, A. (2016). Employee recognition and performance: a field experiment. *Management Science*, *62*(11), 3085–99.

6 Handgraaf, M.J., De Jeude, M.A.V.L., & Appelt, K.C. (2013). Public

praise vs. private pay: effects of rewards on energy conservation in the workplace. *Ecological Economics*, *86*, 86–92.

7 The poll was run by OnePoll on behalf of the company Bonusly. Cariaga, V. (2022). Nearly Half of Americans Quit Their Jobs Because They Feel Unappreciated by Management. Yahoo! Finance: https://finance.yahoo.com/news/nearly-half-americans-quit-jobs-145250545.html

8 Schembra, C. (2021). Gratitude may be the secret to overcoming the talent crisis. Fast Company: https://www.fastcompany.com/90665927/gratitude-may-be-the-secret-to-overcoming-the-talent-crisis

9 Izuma, K., Saito, D.N., & Sadato, N. (2008). Processing of social and monetary rewards in the human striatum. *Neuron*, *58*(2), 284–94.

10 Grant, N.K., Fabrigar, L.R., & Lim, H. (2010). Exploring the efficacy of compliments as a tactic for securing compliance. *Basic and Applied Social Psychology*, *32*(3), 226–33.

11 Grant, N.K., Krieger, L.R., Nemirov, H., Fabrigar, L.R., & Norris, M.E. (2022). I'll scratch your back if you give me a compliment: exploring psychological mechanisms underlying compliments' effects on compliance. *British Journal of Social Psychology*, *61*(1), 37–54.

12 Stsiampkouskaya, K., Joinson, A., & Piwek, L. (2023). To Like or Not to Like? An Experimental Study on Relational Closeness, Social Grooming, Reciprocity, and Emotions in Social Media Liking. *Journal of Computer-Mediated Communication*, *28*(2), zmac036. Available online at: https://doi.org/10.1093/jcmc/zmac036

13 Chan, E., & Sengupta, J. (2010). Insincere flattery actually works: a dual attitudes perspective. *Journal of Marketing Research*, *47*(1), 122–33.

14 Gordon, A.M., & Diamond, E. (2023). Feeling understood and appreciated in relationships: where do these perceptions come from and why do they matter?. *Current Opinion in Psychology*, 101687.

15 See Survey 5 in the Supplemental Materials of Zhao, X., & Epley, N. (2021). Insufficiently complimentary?: Underestimating the positive impact of compliments creates a barrier to expressing them. *Journal of Personality and Social Psychology*, *121*(2), 239.

16 Boothby, E.J., & Bohns, V.K. (2021). Why a simple act of kindness is not as simple as it seems: underestimating the positive impact of our compliments on others. *Personality and Social Psychology Bulletin*, *47*(5), 826–40.

17 Zhao, X., & Epley, N. (2021). Insufficiently complimentary?

18 Zhao, X., & Epley, N. (2021). Kind words do not become tired words: undervaluing the positive impact of frequent compliments. *Self and Identity*, *20*(1), 25–46.

19 Johnson, S., *The Rambler*, no. 136. Saturday, 6 July 1751. Available here: http://www.yalejohnson.com/frontend/sda_viewer?n=106855
20 Kumar, A., & Epley, N. (2018). Undervaluing gratitude: expressers misunderstand the consequences of showing appreciation. *Psychological Science, 29*(9), 1423–35.
21 Toepfer, S.M., Cichy, K., & Peters, P. (2012). Letters of gratitude: further evidence for author benefits. *Journal of Happiness Studies, 13*, 187–201.
22 Gu, Y., Ocampo, J.M., Algoe, S.B., & Oveis, C. (2022). Gratitude expressions improve teammates' cardiovascular stress responses. *Journal of Experimental Psychology: General, 151*(12), 3281–91.
23 Algoe, S.B., & Zhaoyang, R. (2016). Positive psychology in context: effects of expressing gratitude in ongoing relationships depend on perceptions of enactor responsiveness. *Journal of Positive Psychology, 11*(4), 399–415.
24 Kaczmarek, L.D., Kashdan, T.B., Drążkowski, D., Enko, J., Kosakowski, M., Szäefer, A., & Bujacz, A. (2015). Why do people prefer gratitude journaling over gratitude letters? The influence of individual differences in motivation and personality on web-based interventions. *Personality and Individual Differences, 75*, 1–6.
25 Algoe, S.B., Dwyer, P.C., Younge, A., & Oveis, C. (2020). A new perspective on the social functions of emotions: gratitude and the witnessing effect. *Journal of Personality and Social Psychology, 119*(1), 40.
26 Algoe, S.B. (2012). Find, remind, and bind: the functions of gratitude in everyday relationships. *Social and Personality Psychology Compass, 6*(6), 455–69; Algoe, S.B., Dwyer, P.C., Younge, A., & Oveis, C. (2020). A new perspective on the social functions of emotions. See also: https://www.psychologicalscience.org/news/why-saying-thank-you-makes-a-difference.html
27 Algoe, S.B., Kurtz, L.E., & Hilaire, N.M. (2016). Putting the 'you' in 'thank you': examining other-praising behavior as the active relational ingredient in expressed gratitude. *Social Psychological and Personality Science, 7*(7), 658–66.
28 Algoe, S.B., Fredrickson, B.L., & Gable, S.L. (2013). The social functions of the emotion of gratitude via expression. *Emotion, 13*(4), 605; Algoe, S.B., & Zhaoyang, R. (2016). Positive psychology in context: effects of expressing gratitude in ongoing relationships depend on perceptions of enactor responsiveness. *Journal of Positive Psychology, 11*(4), 399–415.
29 Czopp, A.M. (2008). When is a compliment not a compliment? Evaluating expressions of positive stereotypes. *Journal of Experimental Social Psychology, 44*(2), 413–20.

30 Ashokkumar, A., & Swann Jr, W.B. (2020). The saboteur within. In Brummelman, E. (ed.), *Psychological Perspectives on Praise* (pp. 11–18) Routledge; Kille, D.R., Eibach, R.P., Wood, J.V., & Holmes, J.G. (2017). Who can't take a compliment? The role of construal level and self-esteem in accepting positive feedback from close others. *Journal of Experimental Social Psychology*, 68, 40–9.
31 Higgins, E.T. (2019). *Shared Reality* (p. 172). Oxford University Press.
32 Marigold, D.C., Holmes, J.G., & Ross, M. (2007). More than words: reframing compliments from romantic partners fosters security in low self-esteem individuals. *Journal of Personality and Social Psychology*, 92(2), 232. See also Marigold, D.C., Holmes, J.G., & Ross, M. (2010). Fostering relationship resilience: an intervention for low self-esteem individuals. *Journal of Experimental Social Psychology*, 46(4), 624–30.
33 Sezer, O., Prinsloo, E., Brooks, A., & Norton, M.I. (2019). Backhanded compliments: how negative comparisons undermine flattery. https://papers.ssrn.com/sol3/papers.cfm?abstract_id=3439774
34 Williams, L.A., & Bartlett, M.Y. (2015). Warm thanks: gratitude expression facilitates social affiliation in new relationships via perceived warmth. *Emotion*, *15*(1), 1.

CHAPTER 7: *Truth, Lies and Secrets*

1 Schopenhauer, A. (1903). *The Basis of Morality* (A.B. Bullock, Trans.) (p. 70). Swan Sonnenschein & Co. Originally published in 1840.
2 Wicks, R. (2021). Arthur Schopenhauer. *Stanford Encyclopedia of Philosophy*: https://plato.stanford.edu/entries/schopenhauer/
3 This detail can be found in the Rai documentary about Schopenhauer, written by the philosopher Simona Menicocci: https://www.raiplaysound.it/audio/2020/05/Maturadio-Podcast-di-folosofia-Schopenhauer-2bf6db6e-e44b-4549-b122-4d188f52d30a.html
4 Cartwright, D.E. (2010). *Schopenhauer: A Biography*. Cambridge University Press.
5 Schopenhauer, A. (1951). *Essays from the Parerga and Paralipomena* (T. Bailey Saunders, Trans.). The porcupine's tale appears on pages 84–5 of 'Studies in Pessimism'. Available online at https://archive.org/details/in.gov.ignca.17417
6 Gide, A. (2011). *Autumn Leaves* (E. Pell, Trans.), (p. 19). Philosophical Library/Open Road. Kindle Edition.
7 Slepian, M.L., & Koch, A. (2021). Identifying the dimensions of secrets to reduce their harms. *Journal of Personality and Social Psychology*, *120*(6), 1431.

8 Slepian, M.L., Chun, J.S., & Mason, M.F. (2017). The experience of secrecy. *Journal of Personality and Social Psychology*, *113*(1), 1.
9 Slepian, M.L., Kirby, J.N., & Kalokerinos, E.K. (2020). Shame, guilt, and secrets on the mind. *Emotion*, 20(2), 323.
10 This is an Italian saying: *La lingua batte dove il dente duole.*
11 Slepian, M.L., Masicampo, E.J., Toosi, N.R., & Ambady, N. (2012). The physical burdens of secrecy. *Journal of Experimental Psychology: General*, *141*(4), 619.
12 Slepian, M.L., Halevy, N., & Galinsky, A.D. (2019). The solitude of secrecy: thinking about secrets evokes goal conflict and feelings of fatigue. *Personality and Social Psychology Bulletin*, *45*(7), 1129–51.
13 Slepian, M.L., Chun, J.S., & Mason, M.F. The experience of secrecy.
14 Slepian, M.L., Masicampo, E.J., & Ambady, N. (2014). Relieving the burdens of secrecy: revealing secrets influences judgments of hill slant and distance. *Social Psychological and Personality Science*, *5*(3), 293–300.
15 Smith, M. (2022). 25 years after her death, Princess Diana is more popular than King Charles, and the monarchy. YouGov: https://yougov.co.uk/politics/articles/44509-25-years-after-her-death-princess-diana-more-popul
16 Brown, T. (2017). *The Diana Chronicles* (p. 354). Random House. Kindle Edition.
17 Savitsky, K., Epley, N., & Gilovich, T. (2001). Do others judge us as harshly as we think? Overestimating the impact of our failures, shortcomings, and mishaps. *Journal of Personality and Social Psychology*, *81*(1), 44.
18 Gromet, D.M., & Pronin, E. (2009). What were you worried about? Actors' concerns about revealing fears and insecurities relative to observers' reactions. *Self and Identity*, *8*(4), 342–64.
19 Bruk, A., Scholl, S.G., & Bless, H. (2018). Beautiful mess effect: self–other differences in evaluation of showing vulnerability. *Journal of Personality and Social Psychology*, *115*(2), 192; Jaffé, M.E., Douneva, M., & Albath, E.A. (2023). Secretive and close? How sharing secrets may impact perceptions of distance. *PloS One*, *18*(4), e0282643. See also: Smith, E.E. (2019). Your flaws are probably more attractive than you think they are. *Atlantic*: https://www.theatlantic.com/health/archive/2019/01/beautiful-mess-vulnerability/579892
20 Jiang, L., John, L.K., Boghrati, R., & Kouchaki, M. (2022). Fostering perceptions of authenticity via sensitive self-disclosure. *Journal of Experimental Psychology: Applied*, *28*(4), 898.

21 Cited in: Bruk, A., Scholl, S.G., & Bless, H. Beautiful mess effect.
22 John, L.K., Barasz, K., & Norton, M.I. (2016). Hiding personal information reveals the worst. *Proceedings of the National Academy of Sciences, 113*(4), 954–9.
23 Rogers, T., & Norton, M.I. (2011). The artful dodger: answering the wrong question the right way. *Journal of Experimental Psychology: Applied, 17*(2), 139.
24 Gerdeman, D. (2016). How The 2016 Presidential Candidates Misled Us With Truthful Statements. Harvard Business School: https://hbswk.hbs.edu/item/paltering-in-action
25 Rogers, T., Zeckhauser, R., Gino, F., Norton, M.I., & Schweitzer, M.E. (2017). Artful paltering: the risks and rewards of using truthful statements to mislead others. *Journal of Personality and Social Psychology, 112*(3), 456.
26 Trend Watch (2018). Hicks: I Tell 'White Lies': Lookups rise 6500% after Hicks resignation. Merriam-Webster: https://www.merriam-webster.com/news-trend-watch/hicks-i-tell-white-lies-20180228
27 Levine, E.E., & Cohen, T.R. (2018). You can handle the truth: mispredicting the consequences of honest communication. *Journal of Experimental Psychology: General, 147*(9), 1400.
28 Levine, E.E., Roberts, A.R., & Cohen, T.R. (2020). Difficult conversations: navigating the tension between honesty and benevolence. *Current Opinion in Psychology, 31*, 38–43.
29 Levine, Emma E. (2022). Community standards of deception: deception is perceived to be ethical when it prevents unnecessary harm. *Journal of Experimental Psychology: General, 151*(2), 410.
30 Abi-Esber, N., Abel, J. E., Schroeder, J., & Gino, F. (2022). 'Just letting you know…': underestimating others' desire for constructive feedback. *Journal of Personality and Social Psychology*, 123(6), 1362–1385.
31 Henley, A.J., & DiGennaro Reed, F.D. (2015). Should you order the feedback sandwich? Efficacy of feedback sequence and timing. *Journal of Organizational Behavior Management, 35*(3–4), 321–35.
32 Kim, S., Liu, P.J., & Min, K.E. (2021). Reminder avoidance: why people hesitate to disclose their insecurities to friends. *Journal of Personality and Social Psychology, 121*(1), 59.
33 Carter, N.L., & Mark Weber, J. (2010). Not Pollyannas: higher generalized trust predicts lie detection ability. *Social Psychological and Personality Science, 1*(3), 274–9.

CHAPTER 8: *Avoiding Envy and Enjoying Confelicity*

1 Shanahan, M. (2017). Celebrity Humblebrags So Iconic They'll Leave

You Secondhand Embarrassed For Days. Buzzfeed: https://www.buzzfeed.com/morganshanahan/humblebrags-so-bad-theyll-leave-you-secondhand-embarassed

2 Weaver, H. (2017). Meryl Streep's Reaction to the Moonlight Mix-Up Defines the 2017 Oscars. Vanity Fair: https://www.vanityfair.com/style/2017/02/meryl-streep-reaction-to-moonlight-oscar-win

3 Zuo, B. (2023). 'Versailles literature' on WeChat Moments: humblebragging with digital technologies. *Discourse & Communication*, 17504813231164854.

4 Sun, Y. (2020). 'Versailles Literature' Trending on China's Internet: A New Way to Brag. pandaily: https://pandaily.com/versailles-literature-trending-on-chinas-internet-a-new-way-to-brag

5 Schlenker, B.R., & Leary, M.R. (1982). Audiences' reactions to self-enhancing, self-denigrating, and accurate self-presentations. *Journal of Experimental Social Psychology*, *18*(1), 89–104; O'Mara, E.M., Kunz, B.R., Receveur, A., & Corbin, S. (2019). Is self-promotion evaluated more positively if it is accurate? Reexamining the role of accuracy and modesty on the perception of self-promotion. *Self and Identity*, *18*(4), 405–24.

6 Schlenker, B.R. (1975). Self-presentation: managing the impression of consistency when reality interferes with self-enhancement. *Journal of Personality and Social Psychology*, *32*(6), 1030.

7 Zell, E., Strickhouser, J.E., Sedikides, C., & Alicke, M.D. (2020). The better-than-average effect in comparative self-evaluation: a comprehensive review and meta-analysis. *Psychological Bulletin*, *146*(2), 118.

8 Dunning, D. (2011). The Dunning–Kruger effect: on being ignorant of one's own ignorance. In *Advances in Experimental Social Psychology* (Vol. 44, pp. 247–96). Academic Press.

9 Hoorens, V., Pandelaere, M., Oldersma, F., & Sedikides, C. (2012). The hubris hypothesis: you can self-enhance, but you'd better not show it. *Journal of Personality*, *80*(5), 1237–74.

10 Van Damme, C., Hoorens, V., & Sedikides, C. (2016). Why self-enhancement provokes dislike: the hubris hypothesis and the aversiveness of explicit self-superiority claims. *Self and Identity*, *15*(2), 173–90.

11 Van Damme, C., Deschrijver, E., Van Geert, E., & Hoorens, V. (2017). When praising yourself insults others: self-superiority claims provoke aggression. *Personality and Social Psychology Bulletin*, *43*(7), 1008–19.

12 Hoorens, V., Pandelaere, M., Oldersma, F., & Sedikides, C. The hubris hypothesis.

13 Steinmetz, J., Sezer, O., & Sedikides, C. (2017). Impression mismanagement: people as inept self-presenters. *Social and Personality Psychology Compass*, *11*(6), e12321.

14 You can take the full test here: https://openpsychometrics.org/tests/NPI
15 de La Bruyère, J. (1885). *The 'Characters' of Jean de La Bruyère* (H. Van Laun, Trans.) (p. 295). Nimmo.
16 Luo, M., & Hancock, J.T. (2020). Modified self-praise in social media. In Placencia, M.E., & Eslami, Z.R. (eds), *Complimenting Behavior and (Self-) Praise across Social Media: New Contexts and New Insights* (pp. 289–309). Benjamins.
17 Roberts, A.R., Levine, E.E., & Sezer, O. (2020). Hiding success. *Journal of Personality and Social Psychology, 120*(5), 1261–86.
18 Harris, D.I. (2015). Friendship as shared joy in Nietzsche. *Symposium, 19*(1), 199–221.
19 Chan, T., Reese, Z.A., & Ybarra, O. (2021). Better to brag: underestimating the risks of avoiding positive self-disclosures in close relationships. *Journal of Personality, 89*(5), 1044–61.
20 Pagani, A.F., Parise, M., Donato, S., Gable, S.L., & Schoebi, D. (2020). If you shared my happiness, you are part of me: capitalization and the experience of couple identity. *Personality and Social Psychology Bulletin, 46*(2), 258–69.
21 Peters, B.J., Reis, H.T., & Gable, S.L. (2018). Making the good even better: a review and theoretical model of interpersonal capitalization. *Social and Personality Psychology Compass, 12*(7), e12407; Chan, T., Reese, Z.A., & Ybarra, O. Better to brag.
22 Chan, T., Reese, Z.A., & Ybarra, O. Better to brag.
23 Lanyon, C. (2016). Years of Rejection Just Made J.K. Rowling More Determined. *New York Magazine*: https://nymag.com/vindicated/2016/11/years-of-rejection-just-made-j-k-rowling-more-determined.html
24 Brooks, A.W., Huang, K., Abi-Esber, N., Buell, R.W., Huang, L., & Hall, B. (2019). Mitigating malicious envy: why successful individuals should reveal their failures. *Journal of Experimental Psychology: General, 148*(4), 667. See also: Nault, K.A., Sezer, O., & Klein, N. (2023). It's the journey, not just the destination: conveying interpersonal warmth in written introductions. *Organizational Behavior and Human Decision Processes, 177*, 104253.
25 Pascal, B. (1995). *Pensées.* (A.J. Krailsheimer, Trans.) (p. 214). Penguin.

CHAPTER 9: *Asking for Help*

1 Franklin, B. (1906). *The Autobiography of Benjamin Franklin* (pp. 106–7). Houghton, Mifflin & Co. Originally published in 1791.
2 Ryan, A.M., & Shin, H. (2011). Help-seeking tendencies during early adolescence: An examination of motivational correlates and consequences for achievement. *Learning and Instruction, 21*(2), 247–56;

Martín-Arbós, S., Castarlenas, E., & Duenas, J.M. (2021). Help-seeking in an academic context: a systematic review. *Sustainability, 13*(8), 4460.

3 Bamberger, P. (2009). Employee help-seeking: Antecedents, consequences and new insights for future research. *Research in Personnel and Human Resources Management, 28*, 49–98.

4 Moran, J. (2016) *Shrinking Violets* (p. 74). Profile. Kindle Edition

5 Gladwell, M. (2008). *Outliers* (pp. 200–7). Little, Brown.

6 Summaries of these studies can be found in: Bohns, V.K. (2016). (Mis)understanding our influence over others: a review of the underestimation-of-compliance effect. *Current Directions in Psychological Science, 25*(2), 119–23. Some details of the specific methods are quoted from Flynn, F.J., & Lake, V.K. (2008). If you need help, just ask: underestimating compliance with direct requests for help. *Journal of Personality and Social Psychology, 95*(1), 128.

7 Straeter, L., & Exton, J. (2018). Why friends give but do not want to receive money. VoxEU: https://voxeu.org/article/why-friends-give-do-not-want-receive-money

8 Bohns, V.K., Roghanizad, M.M., & Xu, A.Z. (2014). Underestimating our influence over others' unethical behavior and decisions. *Personality and Social Psychology Bulletin, 40*(3), 348–62.

9 See study 2 of the following paper: Whillans, A.V., Dunn, E.W., Sandstrom, G.M., Dickerson, S.S., & Madden, K.M. (2016). Is spending money on others good for your heart? *Health Psychology, 35*(6), 574.

10 For this finding, see study 1 of the above paper. See also: Piferi, R.L., & Lawler, K.A. (2006). Social support and ambulatory blood pressure: an examination of both receiving and giving. *International Journal of Psychophysiology, 62*(2), 328–36.

11 Sneed, R.S., & Cohen, S. (2013). A prospective study of volunteerism and hypertension risk in older adults. *Psychology and Aging, 28*(2), 578.

12 Hui, B.P., Ng, J.C., Berzaghi, E., Cunningham-Amos, L.A., & Kogan, A. (2020). Rewards of kindness? A meta-analysis of the link between prosociality and well-being. *Psychological Bulletin, 146*(12), 1084.

13 Details of these experiments, and Inagaki's theory in general, can be found in the following paper: Inagaki, T.K. (2018). Neural mechanisms of the link between giving social support and health. *Annals of the New York Academy of Sciences, 1428*(1), 33–50. See also: Inagaki, T.K., Haltom, K.E.B., Suzuki, S., Jevtic, I., Hornstein, E., Bower, J.E., & Eisenberger, N.I. (2016). The neurobiology of giving versus receiving support: the role of stress-related and social reward-related neural activity. *Psychosomatic Medicine, 78*(4), 443.

14 Wang, Y., Ge, J., Zhang, H., Wang, H., & Xie, X. (2020). Altruistic

behaviors relieve physical pain. *Proceedings of the National Academy of Sciences*, *117*(2), 950–8; Schreier, H.M., Schonert-Reichl, K.A., & Chen, E. (2013). Effect of volunteering on risk factors for cardiovascular disease in adolescents: a randomized controlled trial. *JAMA Pediatrics*, *167*(4), 327–32.

15 Nakamura, J.S., Kwok, C., Huang, A., Strecher, V.J., Kim, E.S., & Cole, S.W. (2023). Reduced epigenetic age in older adults who volunteer. *Psychoneuroendocrinology*, *148*, 106000. For more information on the 'epigenetic clock', see the following article from the US National Institute of Aging: The epigenetics of aging: What the body's hands of time tell us. https://www.nia.nih.gov/news/epigenetics-aging-what-bodys-hands-time-tell-us

16 Poulin, M.J., Brown, S.L., Dillard, A.J., & Smith, D.M. (2013). Giving to others and the association between stress and mortality. *American Journal of Public Health*, *103*(9), 1649–55.

17 Aknin, L.B., Barrington-Leigh, C.P., Dunn, E.W., Helliwell, J.F., Burns, J., Biswas-Diener, R., . . . & Norton, M.I. (2013). Prosocial spending and well-being: cross-cultural evidence for a psychological universal. *Journal of Personality and Social Psychology*, *104*(4), 635; Hui, B.P., Ng, J.C., Berzaghi, E., Cunningham-Amos, L.A., & Kogan, A. Rewards of kindness?

18 Zhao, X., & Epley, N. (2022). Surprisingly happy to have helped: underestimating prosociality creates a misplaced barrier to asking for help. *Psychological Science*, *33*(10), 1708–31.

19 Jecker, J., & Landy, D. (1969). Liking a person as a function of doing him a favour. *Human Relations*, *22*(4), 371–8.

20 The descriptions of *amae*'s definition and connotations in Japanese, along with the results of these experiments, can be found in the following paper: Niiya, Y., Ellsworth, P.C., & Yamaguchi, S. (2006). *Amae* in Japan and the United States: An exploration of a 'culturally unique' emotion. *Emotion*, *6*(2), 279. See also: Niiya, Y., & Ellsworth, P.C. (2012). Acceptability of favor requests in the United States and Japan. *Journal of Cross-Cultural Psychology*, *43*(2), 273–85.

21 Niiya, Y. (2016). Does a favor request increase liking toward the requester? *Journal of Social Psychology*, *156*(2), 211–21.

22 See Inagaki, T.K., & Eisenberger, N.I. (2012). Neural correlates of giving support to a loved one. *Psychosomatic Medicine*, *74*(1), 3–7; Inagaki, T.K. (2018). Neural mechanisms of the link between giving social support and health. *Annals of the New York Academy of Sciences*, *1428*(1), 33–50.

23 Niiya, Y. (2017). Adult's *amae* as a tool for adjustment to a new environment. *Asian Journal of Social Psychology*, *20*(3–4), 238–43.

24 Marshall, T.C. (2012). Attachment and *amae* in Japanese romantic

relationships. *Asian Journal of Social Psychology, 15*(2), 89–100. For a further discussion, see: Niiya, Y. Does a favor request increase liking toward the requester?

25 Gopnik, A. (2011). *The Philosophical Baby* (p. 243). Random House. Kindle Edition.

26 Bohns, V. (2021). *You Have More Influence Than You Think* (pp. 125–56). WW Norton & Company.

27 Seneca, L.A. (1917). *Moral Epistles*, Volume 1 (R.M. Gummere, Trans.) (p. 307). The Loeb Classical Library.

28 Roghanizad, M.M., & Bohns, V.K. (2022). Should I ask over zoom, phone, email, or in-person? Communication channel and predicted versus actual compliance. *Social Psychological and Personality Science, 13*(7), 1163–72.

29 Inagaki, T.K. Neural mechanisms of the link between giving social support and health.

CHAPTER 10: *Healing Bad Feelings*

1 Anesko, M., Zacharias, G.W. (eds). (2018). *The Complete Letters of Henry James, Volume 1* (p. 195). University of Nebraska Press.

2 Dungan, J.A., Munguia Gomez, D.M., & Epley, N. (2022). Too reluctant to reach out: receiving social support is more positive than expressers expect. *Psychological Science, 33*(8), 1300–12.

3 Hewitt, R. (2018). Do 'animal Fluids move by Hydraulick laws'?: the politics of the hydraulic theory of emotion. *Lancet Psychiatry, 5*(1), 25–6; Littrell, J. (2008). The status of Freud's legacy on emotional processing: contemporary revisions. *Journal of Human Behavior in the Social Environment, 18*(4), 477–99; Evans, D. (2002). *Emotion: The Science of Sentiment* (pp. 81–2). Oxford University Press; Kross, E. (2021). *Chatter: The Voice in Our Head, Why It Matters, and How to Harness It.* Vermilion.

4 Zech, E. (2000). The effects of the communication of emotional experiences [unpublished doctoral dissertation]. University of Louvain. Available at: http://hdl.handle.net/2078.1/149682. Cited in: Rimé, B. (2009). Emotion elicits the social sharing of emotion: Theory and empirical review. *Emotion Review, 1*(1), 60–85.

5 Barasch, A. (2020). The consequences of sharing. *Current Opinion in Psychology, 31*, 61–6.

6 Vicary, A.M., & Fraley, R.C. (2010). Student reactions to the shootings at Virginia Tech and Northern Illinois University: does sharing grief and support over the Internet affect recovery? *Personality and Social Psychology Bulletin, 36*(11), 1555–63; Seery, M.D., Silver, R.C., Holman, E.A., Ence, W.A., & Chu, T.Q. (2008). Expressing thoughts and

feelings following a collective trauma: immediate responses to 9/11 predict negative outcomes in a national sample. *Journal of Consulting and Clinical Psychology*, 76(4), 657. I first became aware of this fascinating research through Ethan Kross's book *Chatter* (see note 3); Kross's interpretations of these findings have been instrumental for the structuring of this chapter.

7 Bastin, M., Vanhalst, J., Raes, F., & Bijttebier, P. (2018). Co-brooding and co-reflection as differential predictors of depressive symptoms and friendship quality in adolescents: investigating the moderating role of gender. *Journal of Youth and Adolescence*, 47, 1037–51.

8 Horn, A.B., & Maercker, A. (2016). Intra-and interpersonal emotion regulation and adjustment symptoms in couples: the role of co-brooding and co-reappraisal. *BMC Psychology*, 4, 1–11.

9 Barasch, A. The consequences of sharing.

10 Starr, L.R., Huang, M., & Scarpulla, E. (2021). Does it help to talk about it? Co-rumination, internalizing symptoms, and committed action during the COVID-19 global pandemic. *Journal of Contextual Behavioral Science*, 21, 187–95.

11 Alparone, F.R., Pagliaro, S., & Rizzo, I. (2015). The words to tell their own pain: linguistic markers of cognitive reappraisal in mediating benefits of expressive writing. *Journal of Social and Clinical Psychology*, 34(6), 495–507; see also Zheng, L., Lu, Q., & Gan, Y. (2019). Effects of expressive writing and use of cognitive words on meaning making and post-traumatic growth. *Journal of Pacific Rim Psychology*, 13, e5.

12 McAdams, D.P. (2013). The psychological self as actor, agent, and author. *Perspectives on Psychological Science*, 8(3), 272–95.

13 See, for example: Slotter, E.B., & Ward, D.E. (2015). Finding the silver lining: the relative roles of redemptive narratives and cognitive reappraisal in individuals' emotional distress after the end of a romantic relationship. *Journal of Social and Personal Relationships*, 32(6), 737–56.

14 Adler, J.M., Turner, A.F., Brookshier, K.M., Monahan, C., Walder-Biesanz, I., Harmeling, L.H., . . . & Oltmanns, T.F. (2015). Variation in narrative identity is associated with trajectories of mental health over several years. *Journal of Personality and Social Psychology*, 108(3), 476.

15 Mitchell, C., Reese, E., Salmon, K., & Jose, P. (2020). Narrative coherence, psychopathology, and wellbeing: concurrent and longitudinal findings in a mid-adolescent sample. *Journal of Adolescence*, 79, 16–25. I have written about these findings previously for *New Scientist*: How to take control of your self-narrative for a better, happier life. https://

www.newscientist.com/article/mg25634204-800-how-to-take-control-of-your-self-narrative-for-a-better-happier-life

16 Mitchell, C., & Reese, E. (2022). Growing memories: coaching mothers in elaborative reminiscing with toddlers benefits adolescents' turning-point narratives and wellbeing. *Journal of Personality*, 90(6), 887–901.

17 Hiller, R.M., Meiser-Stedman, R., Lobo, S., Creswell, C., Fearon, P., Ehlers, A., . . . & Halligan, S.L. (2018). A longitudinal investigation of the role of parental responses in predicting children's post-traumatic distress. *Journal of Child Psychology and Psychiatry*, 59(7), 781–9.

18 Noel, M., Pavlova, M., Lund, T., Jordan, A., Chorney, J., Rasic, N., . . . & Graham, S. (2019). The role of narrative in the development of children's pain memories: influences of father– and mother–child reminiscing on children's recall of pain. *Pain*, 160(8), 1866–75.

19 Pavlova, M., Lund, T., Nania, C., Kennedy, M., Graham, S., & Noel, M. (2022). Reframe the pain: A randomized controlled trial of a parent-led memory-reframing intervention. *Journal of Pain*, 23(2), 263–75.

20 Danoff-Burg, S., Mosher, C.E., Seawell, A.H., & Agee, J.D. (2010). Does narrative writing instruction enhance the benefits of expressive writing? *Anxiety, Stress, & Coping*, 23(3), 341–52.

21 Kross, E., & Ayduk, O. (2017). Self-distancing: theory, research, and current directions. In *Advances in Experimental Social Psychology* (Vol. 55, pp. 81–136). Academic Press. See also: Rude, S.S., Mazzetti, F.A., Pal, H., & Stauble, M.R. (2011). Social rejection: how best to think about it? *Cognitive Therapy and Research*, 35, 209–16.

22 Lee, D.S., Orvell, A., Briskin, J., Shrapnell, T., Gelman, S.A., Ayduk, O., . . . & Kross, E. (2020). When chatting about negative experiences helps – and when it hurts: distinguishing adaptive versus maladaptive social support in computer-mediated communication. *Emotion*, 20(3), 368. For further evidence of the benefits of reconstrual, see: Nils, F., & Rimé, B. (2012). Beyond the myth of venting: social sharing modes determine the benefits of emotional disclosure. *European Journal of Social Psychology*, 42(6), 672–81.

23 Kross, E. *Chatter* (p. 94).

24 Kil, H., Allen, M.P., Taing, J., & Mageau, G.A. (2022). Autonomy support in disclosure and privacy maintenance regulation within romantic relationships. *Personal Relationships*, 29(2), 305–31. You can read the researchers' description of their work on the Character and Context blog of the Society for Personality and Social Psychology: https://spsp.org/news/character-and-context-blog/kil-mageau-allen-open-conversations-with-partner

25 Sanchez, M., Haynes, A., Parada, J.C., & Demir, M. (2020). Friendship maintenance mediates the relationship between compassion for others and happiness. *Current Psychology*, *39*, 581–92.

26 Matos, M., McEwan, K., Kanovský, M., Halamová, J., Steindl, S.R., Ferreira, N., . . . & Gilbert, P. (2021). Fears of compassion magnify the harmful effects of threat of COVID-19 on mental health and social safeness across 21 countries. *Clinical Psychology & Psychotherapy*, *28*(6), 1317–33; Matos, M., McEwan, K., Kanovský, M., Halamová, J., Steindl, S.R., Ferreira, N., . . . & Gilbert, P. (2022). Compassion protects mental health and social safeness during the COVID-19 pandemic across 21 countries. *Mindfulness*, *13*(4), 863–80. See also: Svoboda, E. (2021). Is Avoiding Other People's Suffering Good for Your Mental Health?. *Greater Good Magazine*: https://greatergood.berkeley.edu/article/item/is_avoiding_other_peoples_suffering_good_for_your_mental_health

27 Jazaieri, H., Jinpa, G.T., McGonigal, K., Rosenberg, E.L., Finkelstein, J., Simon-Thomas, E., . . . & Goldin, P.R. (2013). Enhancing compassion: a randomized controlled trial of a compassion cultivation training program. *Journal of Happiness Studies*, *14*, 1113–26.

28 Anesko, M., Zacharias, G.W. (eds). *The Complete Letters of Henry James, Volume 1* (p. 197).

29 Tursi, R. (2017). Cambridge's Grace Norton: an absent presence. *Massachusetts Historical Review*, *19*, 117–48.

CHAPTER 11: *Constructive Disagreement*

1 De Vogue, A. (2016). Scalia-Ginsburg friendship bridged opposing ideologies. CNN: https://edition.cnn.com/2016/02/14/politics/antonin-scalia-ruth-bader-ginsburg-friends/index.html

2 All Things Considered. (2016). Ginsburg and Scalia: 'Best buddies'. NPR: https://www.npr.org/2016/02/15/466848775/scalia-ginsburg-opera-commemorates-sparring-supreme-court-friendship

3 All Things Considered. Ginsburg and Scalia: 'Best buddies'.

4 Cox, C. (2020). Fact check: It's true, Ginsburg and Scalia were close friends despite ideological differences. *USA Today*: https://eu.usatoday.com/story/news/factcheck/2020/09/27/fact-check-ruth-bader-ginsburg-antonin-scalia-were-close-friends/3518592001

5 Public Information Office. (2016). Ruth Bader Ginsberg, Remarks for the Second Court Judicial Conference: https://www.supremecourt.gov/publicinfo/speeches/remarks%20for%20the%20second%20circuit%20judicial%20conference%20may%2025%202016.pdf

6 *USA Today*. (2016). Supreme Court justices weigh in on Antonin Scalia's death. https://eu.usatoday.com/story/news/politics/2016/02/14/statements-supreme-court-death-justice-scalia/80375976
7 Senior, J. (2010). The Ginsburg-Scalia act was not a farce. *New York Times*: https://www.nytimes.com/2020/09/22/opinion/ruth-bader-ginsburg-antonin-scalia.html
8 Green, M. (2020). Why friendships are falling apart over politics. The Conversation: https://theconversation.com/why-friendships-are-falling-apart-over-politics-146821; Pew Research Center. (2019). Partisan antipathy: More intense, More personal. https://www.pewresearch.org/politics/2019/10/10/partisan-antipathy-more-intense-more-personal; Pew Research Center. (2022). As partisan hostility grows, signs of frustration with the two-party system. https://www.pewresearch.org/politics/2022/08/09/as-partisan-hostility-grows-signs-of-frustration-with-the-two-party-system
9 All Things Considered. (2020). 'Dude, I'm done': When politics tears families and friendships apart. NPR: https://www.npr.org/2020/10/27/928209548/dude-i-m-done-when-politics-tears-families-and-friendships-apart
10 Asch, S.E. (1951). Effects of group pressure upon the modification and distortion of judgments. In H.S. Guetzkow (ed.), *Groups, Leadership and Men: Research in Human Relations* (pp. 222–36). Carnegie Press.
11 Friend, R., Rafferty, Y., & Bramel, D. (1990). A puzzling misinterpretation of the Asch 'conformity' study. *European Journal of Social Psychology*, 20(1), 29–44.
12 Gilchrist, A. (2015). Perception and the social psychology of 'The Dress'. *Perception*, 44(3), 229–31.
13 Higgins, E.T. (2019). *Shared Reality* (pp. 150–7). Oxford University Press. Kindle Edition.
14 Pinel, E.C., Long, A.E., & Crimin, L.A. (2010). I-sharing and a classic conformity paradigm. *Social Cognition*, *28*(3), 277–89.
15 Pinel, E.C., Long, A.E., & Crimin, L.A. I-sharing and a classic conformity paradigm.; see also Pinel, E.C., Long, A.E., Murdoch, E.Q., & Helm, P. (2017). A prisoner of one's own mind: identifying and understanding existential isolation. *Personality and Individual Differences*, *105*, 54–63.
16 Graeupner, D., & Coman, A. (2017). The dark side of meaning-making: How social exclusion leads to superstitious thinking. *Journal of Experimental Social Psychology*, 69, 218–22; Poon, K.T., Chen, Z., & Wong, W.Y. (2020). Beliefs in conspiracy theories following ostracism. *Personality and Social Psychology Bulletin*, 46(8), 1234–46.

17 A summary of the survey results can be found on the Pew Research Center's website: https://www.pewresearch.org/global/wp-content/uploads/sites/2/2021/10/PG_2021.10.13_Diversity_Topline.pdf
18 Pinel, E.C., Fuchs, N.A., & Benjamin, S. (2022). I-sharing across the aisle: can shared subjective experience bridge the political divide? *Journal of Applied Social Psychology*, *52*(6), 407–13.
19 Montagu, M.W. (1837). *The Letters and Works of Lady Mary Wortley Montagu Volume III* (p. 134). Richard Bentley.
20 Frimer, J.A., & Skitka, L.J. (2018). The Montagu Principle: incivility decreases politicians' public approval, even with their political base. *Journal of Personality and Social Psychology*, *115*(5), 845.
21 Hessan, D. (2016). Understanding the undecided voters. *Boston Globe*: https://www.bostonglobe.com/opinion/2016/11/21/understanding-undecided-voters/9EjNHVkt99b4re2VAB8ziI/story.html
22 Chen, F.S., Minson, J.A., & Tormala, Z.L. (2010). Tell me more: the effects of expressed interest on receptiveness during dialog. *Journal of Experimental Social Psychology*, *46*(5), 850–3.
23 Yeomans, M., Minson, J., Collins, H., Chen, F., & Gino, F. (2020). Conversational receptiveness: improving engagement with opposing views. *Organizational Behavior and Human Decision Processes*, *160*, 131–48.
24 Itzchakov, G., & Reis, H.T. (2021). Perceived responsiveness increases tolerance of attitude ambivalence and enhances intentions to behave in an open-minded manner. *Personality and Social Psychology Bulletin*, *47*(3), 468–85; Reis, H.T., Lee, K.Y., O'Keefe, S.D., & Clark, M.S. (2018). Perceived partner responsiveness promotes intellectual humility. *Journal of Experimental Social Psychology*, *79*, 21–33.
25 Besides detailing these studies, Itzchakov's paper offers a meta-analysis of the results, which confirms a substantial effect size. Itzchakov, G., Weinstein, N., Legate, N., & Amar, M. (2020). Can high quality listening predict lower speakers' prejudiced attitudes? *Journal of Experimental Social Psychology*, *91*, 104022. For further evidence and detailed discussions of this research, see Itzchakov, G., Reis, H.T., & Weinstein, N. (2022). How to foster perceived partner responsiveness: high-quality listening is key. *Social and Personality Psychology Compass*, *16*(1), e12648.
26 Livingstone, A.G., Fernández Rodríguez, L., & Rothers, A. (2020). 'They just don't understand us': the role of felt understanding in intergroup relations. *Journal of Personality and Social Psychology*, *119*(3), 633.
27 Tucholsky, K. (1932). *Lerne lachen ohne zu weinen* (p. 148). Rowohlt.

28 Chang, C.H., Nastase, S.A., & Hasson, U. (2023). How a speaker herds the audience: Multi-brain neural convergence over time during naturalistic storytelling. *bioRxiv*. Available online at: https://www.ncbi.nlm.nih.gov/pmc/articles/PMC10592711/
29 See the following for a discussion of these psychological mechanisms: Van Bavel, J.J., Reinero, D.A., Spring, V., Harris, E.A., & Duke, A. (2021). Speaking my truth: why personal experiences can bridge divides but mislead. *Proceedings of the National Academy of Sciences*, *118*(8), e2100280118.
30 Kubin, E., Puryear, C., Schein, C., & Gray, K. (2021). Personal experiences bridge moral and political divides better than facts. *Proceedings of the National Academy of Sciences*, *118*(6), e2008389118.
31 Kalla, J.L., & Broockman, D.E. (2020). Reducing exclusionary attitudes through interpersonal conversation: evidence from three field experiments. *American Political Science Review*, *114*(2), 410–25.
32 Kubin, E., Puryear, C., Schein, C., & Gray, K. (2021). Personal experiences bridge moral and political divides better than facts.
33 Feygina, I., Jost, J.T., & Goldsmith, R.E. (2010). System justification, the denial of global warming, and the possibility of 'system-sanctioned change'. *Personality and Social Psychology Bulletin*, *36*(3), 326–38.
34 Feinberg, M., & Willer, R. (2015). From gulf to bridge: when do moral arguments facilitate political influence?. *Personality and Social Psychology Bulletin*, *41*(12), 1665–81; Feinberg, M., & Willer, R. (2019). Moral reframing: a technique for effective and persuasive communication across political divides. *Social and Personality Psychology Compass*, *13*(12), e12501.
35 Horgan, J., Altier, M.B., Shortland, N., & Taylor, M. (2017). Walking away: the disengagement and de-radicalization of a violent right-wing extremist. *Behavioral Sciences of Terrorism and Political Aggression*, *9*(2), 63–77.
36 De Vogue, A. Scalia-Ginsburg friendship bridged opposing ideologies.
37 Boden, A., & Slattery, E. (2022). What we can learn from Antonin Scalia and Ruth Bader Ginsburg's friendship. Pacific Legal Foundation: https://pacificlegal.org/antonin-scalia-and-ruth-bader-ginsburgs-friendship

CHAPTER 12: *Finding Forgiveness*

1 The Beatles Bible (2018). John Lennon and Paul McCartney consider appearing on Saturday Night Live. https://www.beatlesbible.com/1976/04/24/john-lennon-paul-mccartney-saturday-night-live-lorne-michaels
2 White, R. *Come Together: Lennon and McCartney in the Seventies* (p. 7). Omnibus Press. Kindle Edition.

3 White, R. *Come Together* (p. 162).

4 Sample, I. (2018). 'Voodoo doll and cannibalism studies triumph at Ig Nobels'. *Guardian*: https://www.theguardian.com/science/2018/sep/14/voodoo-doll-and-cannibalism-studies-triumph-at-ig-nobels

5 Schumann, K., & Walton, G.M. (2022). Rehumanizing the self after victimization: the roles of forgiveness versus revenge. *Journal of Personality and Social Psychology, 122*(3), 469.

6 Rasmussen, K.R., Stackhouse, M., Boon, S.D., Comstock, K., & Ross, R. (2019). Meta-analytic connections between forgiveness and health: the moderating effects of forgiveness-related distinctions. *Psychology & Health, 34*(5), 515–34. See also: Wade, N.G., & Tittler, M.V. (2019). Psychological interventions to promote forgiveness of others: review of empirical evidence. In Worthington Jr, E.L., & Wade, N.G. (eds), *Handbook of Forgiveness* (pp. 255–65). Routledge.

7 Messias, E., Saini, A., Sinato, P., & Welch, S. (2010). Bearing grudges and physical health: relationship to smoking, cardiovascular health and ulcers. *Social Psychiatry and Psychiatric Epidemiology, 45*, 183–7.

8 Howe, D. (2008). Forgive Me?. *Greater Good Magazine*: https://greatergood.berkeley.edu/article/item/forgive_me; see also McNulty, J.K. (2010). Forgiveness increases the likelihood of subsequent partner transgressions in marriage. *Journal of Family Psychology, 24*(6), 787; McNulty, J.K. (2011). The dark side of forgiveness: the tendency to forgive predicts continued psychological and physical aggression in marriage. *Personality and Social Psychology Bulletin, 37*(6), 770–83; Luchies, L.B., Finkel, E.J., McNulty, J.K., & Kumashiro, M. (2010). The doormat effect: when forgiving erodes self-respect and self-concept clarity. *Journal of Personality and Social Psychology, 98*(5), 734.

9 Curzer, H.J. (2012). *Aristotle and the Virtues* (p. 156). Oxford University Press.

10 These questions all come from the forgiveness scale developed by Mark Rye and colleagues at the University of Dayton. Worthington Jr, E.L., Lavelock, C., vanOyen Witvliet, C., Rye, M.S., Tsang, J.A., & Toussaint, L. (2015). Measures of forgiveness: Self-report, physiological, chemical, and behavioral indicators. In *Measures of personality and social psychological constructs* (pp. 474–502). Academic Press.

11 Rye, M.S., Loiacono, D.M., Folck, C.D., Olszewski, B.T., Heim, T.A., & Madia, B.P. (2001). Evaluation of the psychometric properties of two forgiveness scales. *Current Psychology, 20*, 260–77.

12 Wade, N.G., Hoyt, W.T., Kidwell, J.E.M., & Worthington, E.L. (2014). Efficacy of psychotherapeutic interventions to promote

forgiveness: a meta-analysis. *Journal of Consulting and Clinical Psychology*, *82*(1), 154–70.

13 Ho, M.Y., Worthington, E., Cowden, R., Bechara, A.O., Chen, Z.J., Gunatirin, E.Y., . . . & VanderWeele, T. (2023). International REACH forgiveness intervention: a multi-site randomized controlled trial. Available as a preprint at: https://osf.io/8qzgw

14 McNeill, B. (2017). After four decades, Everett Worthington, leading expert on forgiveness, set to retire from VCU's Department of Psychology. VCUnews https://news.vcu.edu/article/After_four_decades_Everett_Worthington_leading_expert_on_forgiveness; Stammer. L. (2001). Complex Workings of Forgiveness. *Los Angeles Times*: https://www.latimes.com/archives/la-xpm-2001-jun-11-mn-9065-story.html

15 Finkel, E.J., Slotter, E.B., Luchies, L.B., Walton, G.M., & Gross, J.J. (2013). A brief intervention to promote conflict reappraisal preserves marital quality over time. *Psychological Science*, 24(8), 1595–1601.

16 Huynh, A.C., Yang, D.Y.J., & Grossmann, I. (2016). The value of prospective reasoning for close relationships. *Social Psychological and Personality Science*, 7(8), 893–902.

17 Schumann, K. (2014). An affirmed self and a better apology: the effect of self-affirmation on transgressors' responses to victims. *Journal of Experimental Social Psychology*, *54*, 89–96. The long-term benefits were reported in Schumann, K., Ritchie, E.G., & Dragotta, A. (2021). Adapted self-affirmation and conflict management in romantic relationships. Available as a preprint at: https://psyarxiv.com/j3hyk

18 Webb, C.E., Rossignac-Milon, M., & Higgins, E.T. (2017). Stepping forward together: could walking facilitate interpersonal conflict resolution?. *American Psychologist*, 72(4), 374.

19 Schumann, K. (2018). The psychology of offering an apology: understanding the barriers to apologizing and how to overcome them. *Current Directions in Psychological Science*, 27(2), 74–8.

20 Leunissen, J.M., De Cremer, D., van Dijke, M., & Reinders Folmer, C.P. (2014). Forecasting errors in the averseness of apologizing. *Social Justice Research*, 27(3), 322–39.

21 Carpenter, T.P., Carlisle, R.D., & Tsang, J.A. (2014). Tipping the scales: conciliatory behavior and the morality of self-forgiveness. *Journal of Positive Psychology*, 9(5), 389–401.

22 Leunissen, J.M., De Cremer, D., & Reinders Folmer, C.P. (2012). An instrumental perspective on apologizing in bargaining: the importance of forgiveness to apologize. *Journal of Economic Psychology*, 33, 215–22; Leunissen, J.M., De Cremer, D., Folmer, C.P.R., & Van Dijke, M.

(2013). The apology mismatch: asymmetries between victim's need for apologies and perpetrator's willingness to apologize. *Journal of Experimental Social Psychology*, *49*(3), 315–24.

23 Bippus, A.M., & Young, S.L. (2020). How to say 'I'm sorry:' ideal apology elements for common interpersonal transgressions. *Western Journal of Communication*, *84*(1), 43–57; see also Schumann, K. (2014). An affirmed self and a better apology: the effect of self-affirmation on transgressors' responses to victims. *Journal of Experimental Social Psychology*, *54*, 89–96.

24 Frantz, C.M., & Bennigson, C. (2005). Better late than early: the influence of timing on apology effectiveness. *Journal of Experimental Social Psychology*, *41*(2), 201–7.

25 Yu, A., Berg, J.M., & Zlatev, J.J. (2021). Emotional acknowledgment: how verbalizing others' emotions fosters interpersonal trust. *Organizational Behavior and Human Decision Processes*, *164*, 116–35.

26 Forster, D.E., Billingsley, J., Burnette, J.L., Lieberman, D., Ohtsubo, Y., & McCullough, M.E. (2021). Experimental evidence that apologies promote forgiveness by communicating relationship value. *Scientific Reports*, *11*(1), 1–14.

27 White, R. *Come Together* (p. 156).

28 White, R. *Come Together* (p. 222).

29 Billups, A. (2014). Paul McCartney Thankful for Repaired Friendship Before John Lennon's Death. *Time*: https://time.com/3622655/paul-mccartney-john-lennon-friendship

CONCLUSION: *The Thirteenth Law of Connection*

1 Quoted in Herrmann, D. (1998). *Helen Keller: A Life* (p. 208). Knopf.

2 See study 2 of Liu, P.J., Rim, S., Min, L., & Min, K.E. (2023). The surprise of reaching out: appreciated more than we think. *Journal of Personality and Social Psychology*, *124*(4), 754.

3 Pink, D. (2022). *The Power of Regret* (pp. 131–3). Canongate.

4 Harrison Warren, T. (2022). We're in a Loneliness Crisis: Another Reason to Get Off Your Smartphone. *New York Times*: https://www.nytimes.com/2022/05/01/opinion/loneliness-connectedness-technology.html

5 *Fault Lines* (2022). A toxic feed: social media and teen mental health. Al Jazeera: https://www.aljazeera.com/program/fault-lines/2022/5/4/a-toxic-feed-social-media-and-teen-mental-health

6 Bahrampour, T. (2021). Teens around the world are lonelier than a decade ago. The reason may be smartphones. *Washington Post*: https://

www.washingtonpost.com/local/social-issues/teens-loneliness-smart-phones/2021/07/20/cde8c866-e84e-11eb-8950-d73b3e93ff7f_story.html

7 Clark, J.L., Algoe, S.B., & Green, M.C. (2018). Social network sites and well-being: the role of social connection. *Current Directions in Psychological Science*, *27*(1), 32–7.

8 Kumar, A., & Epley, N. (2021). It's surprisingly nice to hear you: misunderstanding the impact of communication media can lead to suboptimal choices of how to connect with others. *Journal of Experimental Psychology: General*, *150*(3), 595.

9 Clark, J.L., Algoe, S.B., & Green, M.C. (2018). Social network sites and well-being: the role of social connection.

10 Agnew, C.R., Carter, J.J., & Imami, L. (2022). Forming Meaningful Connections Between Strangers in Virtual Reality: Comparing Affiliative Outcomes by Interaction Modality. *Technology, Mind, and Behavior*. Online only: https://doi.org/10.1037/tmb0000091

11 This quotation can be found in a letter to Radcliffe College students dated 6 April 1896.

INDEX